DISCARD

D1261732

Practical
Semiconductor
Data Book for
Electronic
Engineers
and Technicians

PRENTICE-HALL SERIES IN ELECTRONIC TECHNOLOGY

DR. IRVING L. KOSOW, editor

CHARLES M. THOMSON AND JOSEPH J. GERSHON, consulting editors

PRENTICE-HALL INTERNATIONAL, INC., *London*
PRENTICE-HALL OF AUSTRALIA, PTY. LTD., *Sydney*
PRENTICE-HALL OF CANADA, LTD., *Toronto*
PRENTICE-HALL OF INDIA PRIVATE LTD., *New Delhi*
PRENTICE-HALL OF JAPAN, INC., *Tokyo*

Practical Semiconductor Data Book for Electronic Engineers and Technicians

John D. Lenk

Consulting Technical Writer

PRENTICE-HALL, INC.

Englewood Cliffs, N.J.

© 1970 by
PRENTICE-HALL, INC.
Englewood Cliffs, N.J.

All rights reserved. No part of this
book may be reproduced in any form
or by any means without permission
in writing from the publisher.

Current printing (last digit):

10   9   8   7   6   5   4   3   2   1

13-693788-8
Library of Congress Catalog Card Number: 70-100085
Printed in the United States of America

*Dedicated to my wife Irene
and daughter Karen*

Preface

There is no lack of data on the subject of semiconductors. There are dozens of books filled to the brim with basic theory and circuits. This great variety, in itself, creates a problem. Where and how do you pin down specific information on a particular phase of semiconductors?

Where, for example, do you find the equation for calculating small-signal gain of a transistor, or the procedure for reverse current measurement of a rectifier diode, or the gate characteristics of a silicon controlled rectifier? Any of these problems would involve thumbing through the contents of several books. The problem is even further complicated if you are not exactly sure what is meant by "small-signal gain" of a transistor. In that circumstance you must consult at least two books—one to find the equation—and another for a concise description of small signal gain.

Although this is not intended as a basic "theory book", one of its major purposes is to bridge the gap between fundamentals of semiconductor theory, and the practical world of the professional technician. This is done by clarifying each equation or procedure with an explanation or introduction. These explanations provide a basic understanding for the student, and serve as a refresher for working technicians.

This book is intended as a companion to the *Data Book for Electronic Technicians and Engineers*. Its major function is to compile semiconductor data that will be of *practical* value to the technician. The book covers a wider range of semiconductor data than any other single book.

There are thousands of semiconductors used in electronics. However, there are only about a dozen different types. Operating characteristics, circuit design procedures (as needed by the technician), and typical applications of each of these types are covered. The last chapter introduces similar data for integrated circuits.

The text also includes basic test procedures for each semiconductor type, as well as an explanation of the data sheet specifications.

So that the condensed information can be put to use by the experimenter, as well as the working technician, the book is rounded out by a summary of the most commonly used semiconductor circuits.

Whether the problem is one of understanding operation of a particular semiconductor type, calculating the component value to meet a specific operating need, pinpointing a particular semiconductor test circuit, or interpreting a data sheet, this book fills the need.

J.D.L.

Contents

CHAPTER 1

Introduction to Semiconductors

This chapter introduces basic solid-state and semiconductor fundamentals from a practical standpoint. The chapter also provides a brief description of the various semiconductor types which are discussed more fully in remaining chapters of the book.

The operation of solid-state devices is related directly to their atomic structure. It is therefore assumed that the reader has a basic understanding of atomic structure, particularly matter, atoms, elements, compounds, molecules, electrons, protons, ions, and the flow of electric current.

1-1. The Basic Semiconductor

Matter or substances can be classified as insulators, conductors, or semiconductors.

Those materials where the electrons are strongly bound to their nuclei are called *insulators*. When a voltage is applied across an insulator, no current will flow since there are no free electrons to support current flow. Best known insulators are rubber, glass, certain ceramic materials, bakelite, and certain plastics.

Those materials where there are many free electrons, and a current will flow easily when a voltage is applied, are called *conductors*. Copper, silver, and some other metals are good conductors of electricity.

In this book, we are concerned with those materials which can not be classified either as insulators or conductors because of their atomic structure.

Such materials are called *semiconductors*. Unlike a conductor, there are no free electrons in a semiconductor. Unlike an insulator, the semiconductor electrons are very weakly attracted by the nucleus. When a voltage is applied to a semiconductor, the electrons overcome the force of attraction produced by the nucleus. Thus, the electrons become "free" and an electric current is produced.

It is difficult to draw the line between most semiconductors and insulators. However, it can be said that a semiconductor is an insulator-type material containing easily dislodgeable electrons. The terms semiconductors and solid-state are often interchanged. Actually, most solid-state devices such as transistors and diodes are semiconductors. The term solid-state is applied since they contain no moving parts (as do switches, relays, etc.) and no filaments or heaters (as do vacuum tubes).

Semiconductor Materials

The most commonly used semiconductor materials are *germanium* and *silicon*. In general, germanium is used for high-frequency operation, while silicon is used in power applications.

The atoms in germanium and silicon contain 4 electrons in their external orbits. These are not free electrons and, when a voltage is applied to *pure* germanium or silicon, very little current will flow. However, if a few atoms of antimony or arsenic are added to the germanium or silicon crystals, the semiconductor will acquire free electrons. Considerable current will flow when a voltage is applied.

When used in manufacture of semiconductors, arsenic and antimony are called *impurities*. If the impurity produces free electrons in the germanium or silicon, the crystals are transformed into *N-type semiconductors*. The semiconductor is considered an *N*-type, or negative, because electric current will flow due to negative charges (electrons). In this case, arsenic or antimony atoms are called *donors*.

Semiconductor Holes

When an atom, normally lacking in free electrons, loses an electron from its outer orbit, the atom acquires a positive charge. The empty space or *hole* in the atom causes the atom to be positively charged, and the atom is no longer neutral. The hole produced by the absence of an electron in a semiconductor has the same charge as that of an electron, but the charge is of opposite polarity. Holes flow in the same manner as electrons, but in the opposite direction.

If a few atoms of aluminum, indium, or gallium are added to the germanium or silicon crystals, the semiconductor will acquire holes. This process is called *doping*. Each hole will behave in a manner similar to that of an electron in an *N*-type semiconductor, except that the hole will move from the positive to the negative terminal of a power source applied to the semiconductor. Such a semiconductor is called a *P*-type since electric current will flow due to positive charges (holes). In this case, the aluminum, gallium, or indium atoms are called *acceptors*.

It is not practical to differentiate between an electron current and a current of holes. However, both types of conduction make possible rectification in crystal diodes, and amplification in transistors.

1-2. The Basic Crystal Junction (or Diode)

When a free electron meets a moving hole in a semiconductor material, the electron occupies the free space, and a positive or negative charge no longer exists. That is, the charge is neutralized. When a *P*-type and an *N*-type crystal are joined to make a single semiconductor, as shown in Fig. 1-1, current will flow in one direction only.

As an example, when a power source is connected to the semiconductor as shown in Fig. 1-2, the semiconductor is said to be *forward biased*. The holes will be repelled toward the junction by the positively charged battery terminal, while the electrons will be pushed toward the junction by the

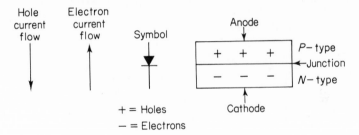

Fig. 1-1. Joining *P*-type and *N*-type crystals.

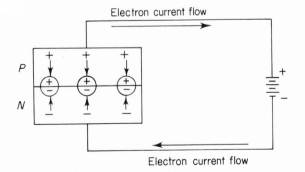

Fig. 1-2. Basic forward-bias connection.

battery's negative terminal. At the junction, the electrons will combine with the holes. Electrons enter the semiconductor at the N-terminal to replace the electrons which have combined with the holes. Likewise, electrons leave the P-terminal by attraction of the positive voltage, and create new holes. This movement of electrons from the negative voltage source through the junction, and from the positive terminal of the semiconductor to the positive voltage source, create a current flow. Thus, current will flow in a semiconductor when the semiconductor is forward biased.

When the polarity of the power source is reversed, and the latter is connected to the semiconductor as shown in Fig. 1-3, the semiconductor is said to be *reverse biased*. The holes are moved away from the junction by the negative voltage, while the electrons are also drawn away from the junction by the positive voltage. Therefore, there is little or no combining of electrons and holes at the junction, and no current will flow. In practical terms, there will always be a few electrons and holes near the junction, allowing a very small current to pass. This small current is known as *leakage current*, and is usually in the order of a few microamperes.

When P-type and N-type regions are formed in the same crystal, the semiconductor is known as a *diode* or *rectifier*. The boundary between the two junctions is termed a *junction*. The P-region terminal is called the *anode* while the N-region terminal is called the *cathode*. Some typical diodes or rectifiers are shown in Fig. 1-4. Usually, when such semiconductors are used with signals, the semiconductors are called *diodes* or *signal diodes*. When the device is used for conversion of alternating current to direct current, the semiconductor is called a rectifier.

Most semiconductor designations are indicated by use of the letter "N," preceded by a number which indicates the number of junctions, and followed by numbers which indicate the order of registration, or an arbitrary assignment. For example, a 1N34 diode indicates a semiconductor with one junction. The 34 which follows the 1N denotes the registration number for the diode.

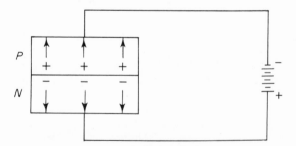

Fig. 1-3. Basic reverse-bias connection.

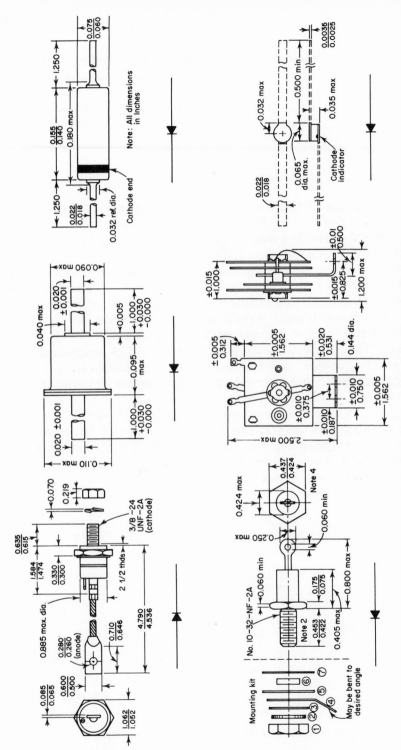

Fig. 1-4. Typical diode assemblies (*courtesy General Electric*).

5

Diodes are often identified by color codes. The standard color codes for diodes, as well as examples of how to use the color codes, are given in Chapter 4.

Diode testing procedures are covered in Chapter 4, while the use of semiconductors in power-supply circuits is discussed in Chapter 5. A special type of *voltage-variable* diode is discussed in Chapter 11.

1-3. The Basic Transistor

Like the diode, a transistor can be used to prevent (or limit) the flow of current in one direction. However, the prime use for a transistor is to control the *amount* of current in a circuit. This is done by adding a second junction to the basic diode junction, discussed in Sec. 1-2.

There are two possible arrangements for the two junctions in transistors: *NPN*, where a positive semiconductor material (holes) is placed between two negative semiconductor materials (electrons), and *PNP*, where a negative material (electrons) is placed between two positive materials (holes).

With either junction arrangement, the basic transistor will have three elements. These elements, shown in Fig. 1-5 as an *NPN* arrangement are: the *emitter* which emits electrons, the *collector* which collects electrons, and a *base* which controls the flow of electrons by controlling the charge concentration at the two junctions on either side of the base.

Figure 1-6 shows how a transistor operates in its basic circuit. As shown, the emitter-base junction will pass current easily because the junction is forward biased. The collector-base junction will not pass current (except for a small leakage current) since the junction is reverse biased. (The term *back bias* is often used in place of reverse bias.)

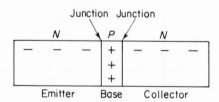

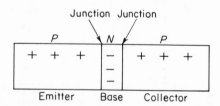

Fig. 1-5. Basic elements for NPN and PNP transistors.

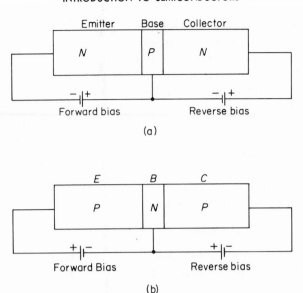

Fig. 1-6. Basic transistor-bias circuits.

It should be noted that the polarities of bias voltages for an *NPN* transistor differ from those of a *PNP* transistor. However, the net results are the same. For example, as shown in (a) of Fig. 1-6, the emitter is negative with respect to the base (*NPN*) to produce a forward bias. In (b) of Fig. 1-6, the emitter is positive with respect to base (*PNP*) to produce the same forward bias. Likewise, the collector has a reverse bias for both *NPN* and *PNP*, even though the polarities are reversed. Also, it should be noted that for normal operation, an *NPN* has its base biased *positively* with respect to its emitter. Conversely, a *PNP* base is *negative* with respect to its emitter.

Basic Transistor Operation

During normal operation of a transistor, current will flow between emitter and base, and between emitter and collector, but not between collector and base. Most of the current flows between emitter and collector because of the large voltage difference existing between these elements (the sum of the emitter-base voltage and the collector-base voltage). This produces a large number of charge carriers (positive holes in a *PNP*, or negative electrons in an *NPN*) which diffuse through the thin base region when passing from emitter to collector (or vice versa). Few of these charge carriers combine with the charges (positive in *NPN*, negative in *PNP*) in the base.

More charge carriers will be pulled out of the emitter, and made available for the collector, if the base-emitter current is increased. This can be accomplished by making the base more negative in a *PNP* transistor, or by making the base more positive in an *NPN* transistor.

If the base-emitter voltage is decreased, less charge carriers will be pulled from the emitter, and less emitter-collector current will flow.

Since very little voltage (approximately 0.2 for germanium and 0.5 for silicon) is required to produce a large current flow in the emitter, input power to a transistor is low. Most of the emitter current will flow in the collector circuit, where the voltage is made much larger. As a result, a relatively large amount of power can be controlled in an external load (connected in the collector circuit), by a small amount of power in the emitter circuit. The power gain of a transistor (the ratio of power output to power input) can be 40,000 or higher in some applications.

Basic NPN Transistor Operation

In an *NPN* transistor, the base-emitter forward bias causes electrons from the emitter to move into the base [Fig. 1-7(a)]. Electrons flow through the

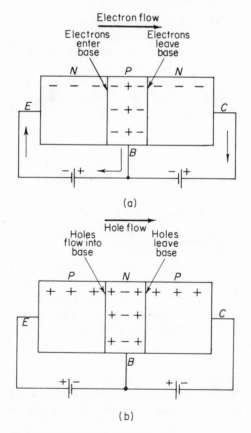

Fig. 1-7. Basic NPN and PNP transistor operation.

thin base into the collector region, with only about 10 per cent of the electrons combining with the base holes. The remaining 90 per cent of the electrons are attracted to the positive collector voltage. In most transistors, the base is kept as thin as possible to maintain the base current at a minimum. Base current is also reduced by keeping the base charges at a minimum. This is done by adding a minimum of impurities to the base material. In effect, base material is only slightly *P* or slightly *N*, in comparison to emitter or collector materials which are heavily doped.

Basic *PNP* Transistor Operation

In a *PNP* transistor [Fig. 1-7(b)], the forward bias between the emitter and base causes holes to flow into the base. Again, the base is relatively thin (and of high purity) so that most of the holes pass through into the collector region. The holes are then attracted to the negative collector voltage, where they are neutralized by electrons from the power source.

Basic Transistor Symbols and Reference Designations

In schematic diagrams, transistors are indicated by appropriate symbols. In rare cases, transistors will be shown by block representations. Figure 1-8 shows the basic transistor symbols, and the corresponding block representations, for both *NPN* and *PNP* transistors.

The only difference between an *NPN* and *PNP* transistor symbol is the arrowhead on the emitter. The arrowhead points away from the base element for an *NPN*, and toward the base for a *PNP*.

Usually, the letters *C* (collector), *E* (emitter) and *B* (base) shown in Fig.

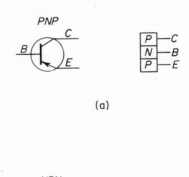

(a)

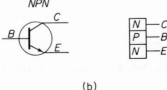

(b)

Fig. 1-8. Basic transistor symbols.

1-8 are omitted from schematic diagrams since the elements can be identified from the symbol.

The appropriate letter portion of a transistor reference designation is Q. Transistors should be identified as Q1, Q2, Q101, Q301, etc. However, transistors can also be identified by the letter "T" or even "V," even though these latter designations are obsolete.

Transistors are covered in much greater detail in remaining chapters of this book as follows: basic circuit data, Chapter 2; testing, Chapter 3; basic design, Chapter 8; typical circuits, Chapter 12; and parameter data, Chapter 13.

1-4. Unijunction Transistors

Unlike the two-junction transistors described in Sec. 1-3, the unijunction (or one junction) transistor is a *negative resistance* unit. Under proper circumstances, the output current can increase, even though the input signal decreases.

The unijunction is formed from an *N*-type silicon material to which another element is bonded, forming a *PN* junction (Fig. 1-9). A unijunction has one emitter lead, two base leads, but no collector lead.

When the voltage applied to the emitter is in reverse-bias form (or zero voltage), the base material acts as a conventional resistor. Because the emitter taps this "resistance," a voltage difference exists between the emitter and ground.

When a forward bias is applied to the emitter, the usual transistor junction characteristics are formed, and the resistance between the emitter and base 1 decreases, causing a current increase. The emitter voltage can then be decreased with a consequent current increase (or negative resistance).

Since unijunction transistors differ from conventional transistors, the operation, circuits, and tests for unijunctions are covered together in Chapter 6.

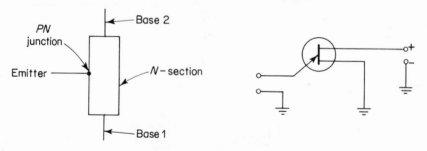

Fig. 1-9. Basic unijunction transistor.

1-5. Field-Effect Transistors

The basic field-effect transistor (or FET) consists of P and N junctions similar to that of conventional transistors. However, the operation of a FET is quite different from that of the basic transistor.

As shown in Fig. 1-10, the primary current flow is produced by battery B_2, and flows through the N section and the load resistor R. In an N-type material, the current flow consists primarily of electrons.

Battery B_1 applies a reverse bias to both P regions, with respect to the N region. The negative battery terminal forms a closed circuit with the P regions through the signal-applying circuit at the input. A sinewave signal applied to the input will increase or decrease the reverse-bias voltage applied by B_1.

When reverse bias is applied to the FET, the bias causes current carriers to move away from the PN junctions, leaving depletion regions. The greater the encroachment of the depletion region on the N region, the less carriers will be present in the N region for current conduction. The input signal can add to, or subtract from, the reverse bias, and thus move the depletion regions more of less into the N region. As a result, the input signal can control the amount of current in the N region.

It is possible to control a large amount of current flow with a small input signal. The large current through the N region also passes through the load resistor, and produces a large voltage drop across the resistor. Thus, a small input voltage variation will produce a large load (or output) voltage variation.

In addition to the depletion FET, there is also an *enhancement*-mode

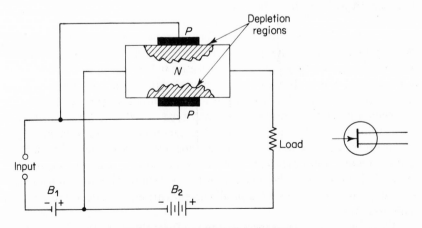

Fig. 1-10. Basic field-effect transistor.

FET. Since both FETs differ from conventional transistors, their operation, typical circuits, and tests are covered in Chapter 7.

1-6. Controlled Rectifiers

With one major exception, the controlled rectifier is similar to the basic diode discussed in Sec. 1-2, and in Chapters 4 and 5. Both the controlled rectifier and the diode will pass current in one direction (from cathode to anode), but not in the opposite direction. The basic diode remains in this condition at all times. The controlled rectifier must be "turned on" by a "gate" or "trigger" signal. Because of this special characteristic the operation, typical circuits, and tests for the various types of controlled rectifiers are covered in Chapter 9.

1-7. Photocells

Photocells provide control of circuits in response to changes in light. There are two basic types of photocells.

The *photovoltaic cell*, or "sun battery" produces a voltage in the presence of light. Both silicon and selenium are used in photovoltaic cells. However, silicon is the most effective.

The *photoconductive cell*, often made of cadmium sulfide, does not generate an electric current. In effect, a photoconductive cell is a *light sensitive resistor*, or *light dependent resistor*.

Data for both types of photocells are covered in Chapter 10.

1-8. Voltage-Variable Diode

All of the elements necessary to form a capacitor (a dielectric between two conductor plates) are present in a *PN* junction diode. The junction between the *P* and *N* regions acts as dielectric (when reverse biased and there is no current flow). The capacitor "plates" are formed by the *P* and *N* regions. In effect, the *PN* junction acts like a slightly charged capacitor.

When a junction is formed between *N*-type and *P*-type material, there is a cross-migration of charges across the junction. Electrons from the *N* region cross the junction to neutralize positive carriers near the junction in the *P* region, and holes from the *P* region cross the junction to neutralize "excess" electrons near the junction in the *N* region. The area immediately around the junction is depleted of all current carriers. In effect, the junction is an insulator.

When the reverse bias is increased to reinforce this condition, the depleted region also increases. This has the same effect as increasing the distance between the P and N regions. As in the case of a conventional capacitor, the capacitance value decreases when the distance between the plates is increased.

If the reverse bias is lowered, the depletion area is also lowered, and the capacitance value increases. Therefore, the value of the junction capacitance is related to the reverse bias voltage. This characteristic can be put to good use when it is desired to have a capacitor that will vary in value with changes in voltage.

Data for all types of voltage-variable diode capacitors are covered in Chapter 11.

1-9. Integrated Circuits

Conventional transistors and diodes are normally used in conjunction with other electronic components (resistors, capacitors, coils, etc.) to form complete circuits. The semiconductor materials used to manufacture transistors and diodes can also be used to produce other circuit components, particularly resistors.

It is possible to form a transistor or diode, and the related circuit components, on the same block of semiconductor material. By using micro-miniaturization techniques, it is possible to produce several transistors and the related components to form a circuit, such as a multi-stage amplifier, on a single block or "chip" of semiconductor material (silicon, germanium, etc). These devices are known as an *integrated circuit*, or IC. They are also referred to as *monolithic* integrated circuits since the parts are made from a single chip of material.

The use of integrated circuits offers the technician or engineer a complete (or nearly complete) functioning circuit, all housed within a container not much larger than the smallest transistor size. Actually, several complete circuits can be contained within a typical transistor housing. However, such a combination is usually of little practical value.

Because of the specialized nature of ICs, they are covered in Chapter 14.

CHAPTER 2

Basic Transistor Circuit Data

This chapter discusses basic transistor circuits from a practical standpoint. The discussion covers such subjects as: transistor types, the three basic transistor circuits, how transistors operate in circuits, how signals are reproduced and amplified by transistors, how gain is produced in an amplifier circuit, and how the necessary bias voltages are applied to transistors.

Practical transistor testing methods are covered in Chapter 3, while technician-level design of transistor circuits is discussed in Chapter 8. Chapter 13 provides a summary of the theoretical approach to transistor characteristics.

2-1. Basic Transistor Types

There are many types of transistors in use today, and new types are being developed constantly. Transistor types can be classified by the method of construction. This is convenient since many of the operating characteristics of a transistor are directly related to its construction.

The following paragraphs describe the basic construction for the most commonly used transistors. The Appendix contains the outline drawings and base diagrams for transistors in general use, as well as some special-purpose transistors, heat sinks, and diodes.

Outer Construction of Transistors

As shown in the Appendix, transistors come in various sizes and shapes. Often, the larger transistors are bolted to the chassis so that the flanges help dissipate the heat generated by higher-power transistors. There are two methods in general use for attaching power transistors to the chassis. In one case, the attaching bolts pass directly through the transistor flanges (which act as heat sinks). In another configuration, a threaded terminal is provided to permit the transistor to be bolted to the chassis.

The smaller type transistors are usually wired directly into the circuits, with the wire leads providing the only mechanical support. In some cases, smaller transistors are plugged into miniature sockets.

When transistors are mounted on insulated boards, rather than on the metal chassis, it is often necessary to provide a separate heat sink for the transistor. (See Appendix.) Heat is dissipated into the surrounding air (rather than into the chassis) by the fins of these heat sinks.

Inner Construction of Junction Transistors

The three regions (*NPN* or *PNP*) of a typical junction transistor are made from the same crystal. Several methods are used to produce junction transistors. The *grown junction* method has been used for some time, and is still in common use. When an *NPN* transistor is to be produced by the grown junction method, a pellet of *P*-forming impurity (such as aluminum, gallium, or indium) is added to molten germanium, into which a single crystal is submerged and pulled out. The *P* impurity melts and sticks to the crystal. Then, *N*-type material is added by re-submerging the crystal in the molten germanium. One problem with the grown junction method is that the thickness of layers is difficult to control. Also, the base consists of a very thin layer, making it difficult to attach a terminal lead.

These problems are overcome in the *alloyed-junction* transistor which is used for most high power applications. The alloyed-junction transistor is produced by melting an impurity pellet to each side of a base. The impurity pellets are of the opposite type material (*P* impurity for an *N* base, and vice

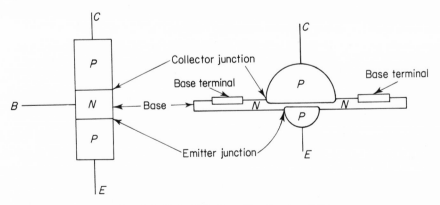

Fig. 2-1. Basic *PNP* alloyed-junction transistor.

versa). The base is a thin wafer of semiconductor material to which is attached a contact or terminal ring, rather than a terminal lead. A *PNP* alloyed-junction transistor is shown in Fig. 2-1.

Special problems are created when transistors are to be operated at higher frequencies. The operating frequency of a transistor is limited by the thickness of the base material, and by the area of the emitter and collector regions. The thicker the base, the longer the transit time for electrons moving between the emitter and collector. When the base thickness is increased to a point where the time consumed in travel between emitter and collector does not permit a change in the direction of current flow, the upper operating frequency limit of the transistor is reached. Also, the larger the emitter and collector regions, the greater the capacity between the two regions. Capacitors tend to store electrons. If the capacitance is high, the storage time is long, and the transistors will not operate efficiently at high frequencies.

Base thickness can be reduced just so far in practical applications. Likewise, a reduction of emitter and collector size produces a corresponding reduction in power handling capabilities of the transistor.

There are several ways to eliminate or minimize these problems. The *surface-barrier* transistor provides satisfactory operation up to about 70 MHz. Surface-barrier transistors are formed by an *electro-chemical* process. As shown in Fig. 2-2, a *PNP* surface-barrier transistor is formed by placing an *N*-type material between two fine streams of an electrolytic solution, and applying a d-c voltage which causes the solution to etch away the semiconductor material until the desired thickness is reached. The polarity of the d-c voltage is then reversed. This causes the metallic solution to deposit small metal dots on each side of the thinned *N* region. The *N*-type material becomes the base, and the dots become the collector and emitter.

The emitter-collector voltage of a surface-barrier transistor is usually about 6 volts maximum, so power capabilities are relatively low.

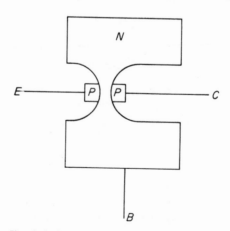

Fig. 2-2. Basic *PNP* surface-barrier transistor.

Where both high frequency and high power are required, the *mesa* transistor, or some form of *planar* transistor, is used effectively. These transistors permit operation at very high frequencies (about 100 MHz for silicon, and 1000 MHz for germanium) with corresponding high power.

As shown in Fig. 2-3(a), the collector of a *mesa* (Spanish word for table or hill) transistor is formed by a layer of *P*-type semiconductor material diffused to a thin layer of *N*-type impurity which becomes the base. An emitter is added by alloying. The table-like structure is produced by etching away the excess material. Since the base area is made small by this etching, the capacity between transistor elements remains low, permitting high frequency operation. However, the large collector will permit use of high voltages (300 volts) and corresponding high power.

An improved version of the mesa transistor is shown in Fig. 2-3(b) where a high-resistance layer of *P*-type material is added before the *N*-type

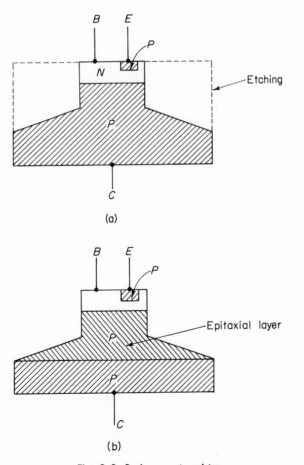

Fig. 2-3. Basic mesa transistor.

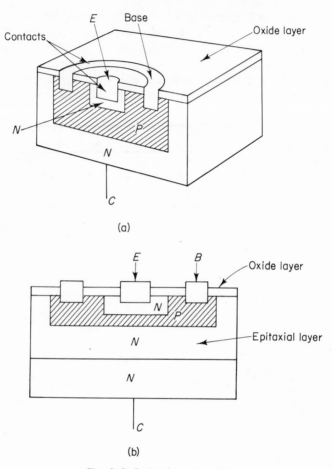

Fig. 2-4. Basic planar transistor.

base is diffused. This high-resistance layer (termed an *epitaxial* layer) allows higher collector voltages to be used.

As shown in Fig. 2-4(a), the collector of a *planar* transistor is formed by a wafer of *N*-type silicon, which has been passivated (coated) with an oxide layer. A circular trench is etched out of the oxide, and filled with a *P*-type base. A disc-shaped area of the oxide is etched at the center, and filled with an *N*-type emitter which is diffused to the base. The oxide layer protects the emitter-base and collector-base junctions from shorts or contamination.

An improved version of the planar transistor is shown in Fig. 2-4(b) where a high-resistance epitaxial layer is added to permit higher collector voltages.

High frequency operation can also be obtained with a *tetrode* transistor (Fig. 2-5). The tetrode operates like a junction transistor, except that there

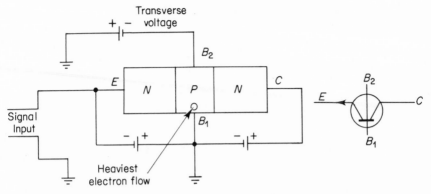

Fig. 2-5. Basic tetrode transistor.

is an additional terminal on the base. A voltage (termed *transverse* voltage) is applied between the two base terminals. This voltage causes the transistor action to occur at one side or the other of the transistor.

In the example of Fig. 2-5, the base is made negative by the transverse voltage. This restricts the flow of electrons between emitter and collector. At the bottom of the base, the transverse voltage is near zero, permitting normal electron flow. Therefore, electron flow is heaviest at the bottom. This has the effect of reducing the junction area, with a corresponding reduction of capacitance. The lower capacitance permits higher frequency operation.

The transverse voltage applied to B_2 (base terminal two) can be varied as necessary to control the effective area of the junction. The transverse voltage is fixed, and will not affect the applied signal impressed between emitter and B_1 (base one).

2-2. Basic Transistor Circuit Connections

There are three basic transistor circuits:

(a) Common-base (or grounded base).
(b) Common-emitter (or grounded emitter).
(c) Common-collector (or grounded collector).

Figure 2-6 shows the three basic methods of connection. As shown, the major difference in the three basic circuits depends on which terminals are used for input and output.

Each configuration in Fig. 2-6 is identified by the transistor element which is *common to both the input and output*. Although it is frequently called the *grounded* element, the common element is not necessarily connected to ground. Figure 2-6 also shows the power supply connections for both *NPN*- and *PNP*-type transistors.

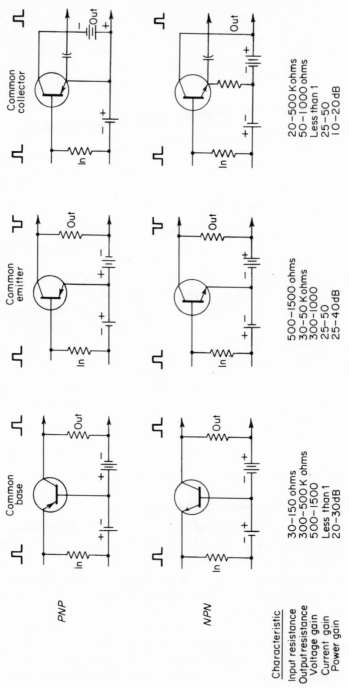

Fig. 2-6. Basic transistor circuit connections.

Characteristic	Common base	Common emitter	Common collector
Input resistance	30–150 ohms	500–1500 ohms	20–500 Kohms
Output resistance	300–500 K ohms	30–50 Kohms	50–1000 ohms
Voltage gain	500–1500	300–1000	Less than 1
Current gain	Less than 1	25–50	25–50
Power gain	20–30dB	25–40dB	10–20dB

In the *common-base* circuit, the input signal is applied between the base and emitter, and the output signal appears between base and collector. Although there is no current gain between input and output of a common-base circuit, it is possible to get considerable *power* gain. This is due to the difference in resistance between input and output circuits.

In the *common-emitter* circuit, the input signal is applied between the base and emitter, and the output signal appears between emitter and collector. This provides a very low-input impedance, and a very high-output impedance. However, the output signal is *out-of-phase* with the input.

Common-emitter circuits are the most often used since there is current gain, resistance gain, and power gain.

In the *common-collector* circuit, the input signal is applied to the base, and the output signal appears at the emitter. This provides a very low-output impedance, and a very high-input impedance. Normally, this condition would produce no gain. However, because the emitter-collector (output) circuit will conduct more current than the base circuit, there is some current gain. The common-collector circuit is also known as an *emitter-follower*. (The output signal is taken from the emitter, and follows the input as to phase.)

Figure 2-6 also shows the similarities and differences in characteristics for the three basic circuits.

The actual input and output impedances of the three circuits will vary with transistor types used (*PNP* or *NPN*), and to some extent with individual units of the same type.

2-3. Operation of Transistors in Circuits

Figure 2-7 shows a *PNP* transistor connected in a basic common-emitter circuit. The base-emitter circuit is forward biased, while the emitter-collector circuit is reverse biased. An input signal is applied across resistor R_1, while the output is taken from across resistor R_2.

Under no-signal (quiescent) conditions, current flows in the input circuit causing a steady value of current to flow in the output circuit. When one alternation of an input signal is applied, as shown in Fig. 2-7(a), a voltage is developed across the input resistor R_1. This voltage, positive at the base end of R_1, subtracts from the bias voltage provided by B_1, causing the base-to-emitter voltage, V_{BE}, to become less negative. Assume that the B_1 battery voltage is 3.0 and the signal voltage at its peak is 0.5 volt. Since the signal voltage drop opposes the B_1 battery voltage, the net voltage between the base and emitter will decrease to 2.5 volts, from the original 3.0. The input current will also decrease, and there will be less current carriers (holes, in this case) available to the collector. Less current will then flow through the output resistor R_2, and the voltage drop across R_2 will decrease.

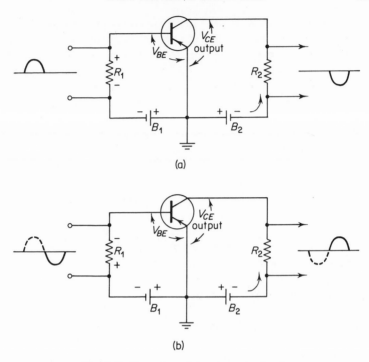

Fig. 2-7. Operation of transistors in circuits.

The voltage from the collector to emitter will increase, the collector becoming more negative than it was during the no-signal condition. It appears that as the signal voltage increases in a positive direction, the output voltage (taken from collector ground) increases in a negative direction. This action illustrates signal-phase reversal, a characteristic of the common-emitter stage. Amplification occurs because the collector current is many times greater than the base-emitter current.

When the next alternation of the input signal is applied [Fig. 2-7(b)], the voltage is again developed across R_1. However, this voltage is negative at the base. The voltage drop adds to the B_1 battery voltage, so that the net voltage between base and emitter will increase to 3.5, from the original 3.0. The input current will increase, and there will be more current carriers (holes) available to the collector. More current will flow through R_2, and the voltage drop across R_2 will increase. This increase will still be out-of-phase with the input, and will be amplified by the same amount as the previous alternation.

The total voltage V_{BE} is a combination of the input signal and B_1, while the total voltage V_{CE} is a combination of output signal and B_2. In practical circuits, the input is measured from base to ground, with the output measured from collector to ground.

If an *NPN* transistor were placed in the same circuit, operation would be identical, except that polarities of the voltages would be reversed.

2-4. Power Gain in Transistors

The fact that power gain exists in a transistor connected as a common-emitter is fairly obvious. Since there is always some current gain between input and output, there is also a corresponding power gain. However, it is also possible to get power gain in a common-base transistor circuit, even though there is no current gain.

Figure 2-8 shows how this power gain in common-base circuits is possible. (The subject of transistor gain is discussed further from a practical standpoint in Chapter 3, and from a theoretical standpoint in Chapter 13.

D-c gain of a transistor is found by dividing the collector current by the emitter current, with a constant collector voltage. However, most transistor data sheets specify some form of a-c or dynamic gain.

The *a-c or dynamic-current gain* (or Alpha) of a common-base transistor is approximately 0.9 to 0.999, and is always less than one. Current gain (dynamic) is calculated by dividing the *difference in collector current* by the *difference in emitter current*, with a constant-collector voltage.

For example, if the emitter current changes by 33 mA, and the collector current changes by 30 amperes, when a signal is applied to the emitter-input circuit, then:

30 divided by 33 = 0.91 and the current gain (Alpha) is 0.91

Resistance gain is calculated by dividing the output (collector) resistance by the input (emitter) resistance.

For example, if the input resistance were 30 ohms and the output resistance were 300,000 ohms, then:

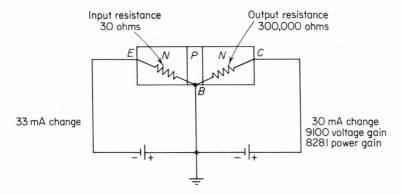

Fig. 2-8. Power gain in transistors.

300,000 divided by 30 = 10,000 ohms resistance gain

Voltage gain is calculated when the current gain is multiplied by the resistance gain, following basic Ohm's law of $E = I \times R$.

For example, if the current gain is 0.91 and the resistance gain is 10,000, then:

0.91 multiplied by 10,000 = 9100 voltage gain

Power gain is calculated when the current gain is multiplied by the voltage gain.

For example, if the current gain is 0.91 and the voltage gain is 9100, then:

0.91 multiplied by 9100 = 8281 power gain

This represents an approximate 39 dB power gain.

It should be noted that while the power gain of a common-emitter circuit is usually greater than that of the common-base circuit (because of the high current gain and some resistance gain), the voltage gain of a common-base circuit is greater (because of the high resistance gain).

2-5. Bias Arrangements for Transistors

In the transistor circuits discussed thus far, two batteries (or power sources) have been shown (one source for emitter-base, and one for emitter-collector). While this arrangement is used, a single battery source is found more practical for many circuits.

Figures 2-9, 2-10, and 2-11 show the bias arrangements (using a single voltage source) for the common-base, common-emitter, and common-collector circuits respectively.

In the *common-base circuit* (Fig. 2-9), the collector-base voltage is taken directly from the battery. Since a *PNP* is shown, reverse bias for the collector base is obtained by connecting the collector negative, with respect to the base. Forward bias for the emitter-base circuit requires that the emitter be positive with respect to the base. This is done with voltage divider resistors R_3 and R_4 connected so that the electron flow from the battery is as shown by the arrow. This current flow causes the correct voltage polarity to appear across R_3, and places the emitter at a positive potential with respect to the base. If an *NPN* transistor were used, the battery would be reversed.

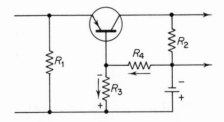

Fig. 2-9. Single source bias for common-base circuit.

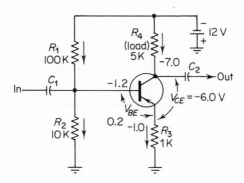

Fig. 2-10. Single source bias for common-emitter circuit.

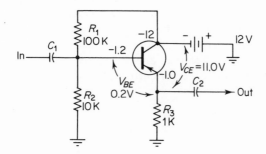

Fig. 2-11. Single source bias for common-collector circuit.

In the *common-emitter circuit* (Fig. 2-10), 12 volts are dropped across R_1 and R_2 to place the base at -1.2 volts (R_1 and R_2 form a 10:1 divider). Assuming that the base-to-emitter voltage (V_{BE}) will be approximately 0.2 volt (typical for germanium), the drop across R_3 would be -0.1 volt. Therefore, the emitter is positive (less negative) with respect to the base, and the base-emitter is forward biased. The voltage from collector-to-emitter (V_{CE}) is typically 6.0 volts. Therefore, the voltage across R_4 would be 5 volts, making the collector-to-ground voltage -7.0 volts ($-12 + 5 = -7$). Therefore, the collector is more negative than the base, and the collector-base junction is reverse biased.

In the *common-collector circuit* (Fig. 2-11), bias operation is essentially the same as for Fig. 2-10, except that the collector is at -12 volts in the common collector circuit.

2-6. Rules for Labeling, Biasing, and Polarities in Transistors

The following general rules can be helpful in a practical analysis of how a transistor circuit will operate. The rules apply primarily to a Class A amplifier, but also remain true for many other transistor circuits.

(a) The middle letter in *PNP* or *NPN* always applies to the *base*.

(b) The first two letters in *PNP* or *NPN* refer to the *relative bias* polarities of the *emitter* with respect to either the base or collector.

For example, the letters *PN*(in *PNP*) indicate that the emitter is positive with respect to both the base and collector. The letters *NP*(in *NPN*) indicate that the emitter is negative with respect to both the base and collector.

(c) The d-c *electron-current flow* is always against the direction of the arrow on the emitter.

(d) If electron flow is into the emitter, electron flow will be out from the collector.

(e) If electron flow is out from the emitter, electron flow will be into the collector.

(f) The collector-base junction is always reverse biased.

(g) The emitter-base junction is always forward biased.

(h) A *base-input* voltage that opposes or decreases the forward bias, also decreases the emitter and collector currents.

(i) A *base-input* voltage that aids or increases the forward bias, also increases the emitter and collector currents.

Practical Transistor Testing

This chapter describes transistor characteristics and test procedures from the practical, technician-level standpoint. That is, the data in this chapter will permit the technician to test all of the *important* transistor characteristics, and to understand the basis for such tests. Chapter 13 provides a summary of the theoretical approach to transistor characteristics (or parameters) that can be used as a supplement to the practical data of this chapter.

3-1. Basic Transistor Tests

Transistors are subjected to a variety of tests during manufacture. It is neither practical nor necessary to duplicate all of these tests in the field. There are only four basic tests required in practical applications: gain, leakage, breakdown, and switching time. Unless a transistor is going to be used for pulse or digital work, the switching characteristics are not of great importance.

In the final analysis, the only true test of a transistor is in the circuit with which the transistor is to be used. However, except in special circumstances, a transistor will operate properly *in-circuit* provided: (1) the transistor shows the proper gain, (2) it does not break down under the maximum operating voltages, (3) the leakage (if any) is within tolerance and, in the case of pulse circuits, the switching characteristics (such as delay time, storage time, etc.) are within tolerance.

There are two exceptions to this rule. Transistor characteristics will change with variations in operating *frequency* and *temperature*. For example, a transistor may be tested at 1 MHz and show more than enough gain to meet circuit requirements. However, at 10 MHz, the gain of the same transistor may be zero. This can be due to a number of factors. Any transistor will have some capacitance at the input and the output. As frequency increases, the capacitive reactance will change until, at some frequency, the transistor will become unsuitable for the circuit. In the case of temperature, the current flow in any junction will increase with increases in temperature. A transistor may be tested for leakage at a normal ambient temperature, and show a leakage well within tolerance. When the same transistor is used "in-circuit," the temperature will increase, increasing the leakage to an unsuitable level.

It is usually not practical to test transistors over the entire range of operating frequencies and temperatures with which the transistor will be used. Instead, the transistor should be tested under the conditions specified in the data sheet. Then, using equations and graphs, the transistor characteristics can be predicted at other frequencies and temperatures. Such equations and graphs, together with an explanation of their use, are given in Chapter 13. The remainder of this chapter is devoted to tests of transistors in relation to data sheet characteristics.

3-2. Leakage Tests

Since *PNP* and *NPN* transistors can be considered as two diodes connected back-to-back, the procedures for transistor leakage tests are similar to those of diodes. In theory, there should be no current flow across a diode junction when the junction is reverse biased. Any current flow under these conditions is the result of current leakage. In the case of a transistor, the collector-base junction is reverse biased, and should show no current flow. However, in practical applications, there will be some collector-base current flow, particularly as the collector voltage is operated near its limits, and as the operating temperature of the transistor is increased.

Collector leakage current is designated as I_{CBO} or I_{CO} on most data sheets. Collector leakage can also be termed "collector-cutoff current" on other data sheets, where a nominal and/or maximum current is specified for a given collector-base voltage and ambient temperature. Collector-base leakage is normally measured with the emitter open, but can also be measured with the emitter shorted to the base, or connected to the base through a resistance.

Figure 3-1 shows the basic circuits for the collector-base leakage test. As shown, the circuits of Fig. 3-1(a), (b), and (c) are for *PNP* transistors, while the circuits of Fig. 3-1(d), (e), and (f) are for *NPN* transistors. Although any

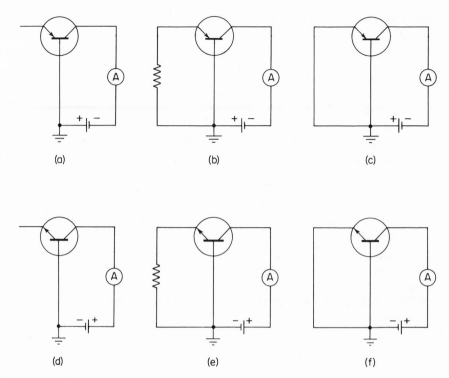

Fig. 3-1. Collector-base leakage test circuits.

of the circuits could be used, those of Fig. 3-1(a) and 3-1(d) (emitter open) are the most popular.

The procedure is the same for all of the circuits shown in Fig. 3-1. The voltage source is adjusted to a given value (thus providing a given reverse bias), and the current (if any) is read on the meter. This current must be below a given maximum, for a given reverse bias.

Temperature is often a critical factor in leakage measurements. Using a 2N332 transistor as an example, the maximum collector leakage current at 25°C is 2 microamperes (with 30 volts applied between collector and base). However, when the collector-base voltage is lowered to 5, and the temperature is raised to 150°C, the maximum collector leakage is 50 microamperes.

Some data sheets also specify emitter-base current leakage. However, this is not usually the case since the emitter-base junction is usually forward-biased in most circuits. Should it become necessary to test the emitter-base current leakage (I_{EO} or I_{EBO}), the circuits of Fig. 3-1 can be used, except that the collector and emitter connections are interchanged. That is, the emitter-base junction is reverse biased, and the collector is left open. The procedures for testing emitter-base leakage are identical to those for collector-base voltage.

CAUTION

Transistors should not be tested for any characteristic unless all of the characteristics are known. The transistor could be damaged if this rule is not followed. Even if no damage results, the test results could be inaccurate. For example, never test a transistor with voltages, or currents, higher than the rated values. The *maximum current rating* is often overlooked. For example, if a transistor is designed to operate with a maximum of 45 volts at the collector, it could be assumed that a 9-volt battery would be safe for all measurements involving the collector. However, assuming that the internal (emitter-to-collector) resistance of the transistor was 90 ohms, and the maximum emitter-collector current was 25 milliamperes, the maximum current would be exceeded. With the 9 volts connected directly between the emitter and collector, the emitter-collector current would be 100 milliamperes, four times the maximum rated 25 milliamperes. This could cause the junctions to overheat and damage the transistor.

3-3. Breakdown Tests

The circuits and procedures for transistor breakdown tests are similar to those for leakage tests. The most important breakdown test is to determine the collector-base breakdown voltage. In this test, the collector and base are reverse biased, while the emitter is open, and the voltage source is adjusted to a *given value of leakage current*. The voltage is then compared with the minimum collector breakdown voltage specified for the transistor. For example, the minimum collector breakdown voltage specified for a 2N332 transistor is 45 volts (with 50 microamperes flowing, and an ambient temperature of 25°C). If 50 microamperes will flow with less than 45 volts, the transistor collector-base junction is breaking down.

Another breakdown test specified on some transistor data sheets is the collector-emitter breakdown voltage. In this test, the collector and emitter are reverse biased, with the base open. The voltage source is then adjusted for a given value of leakage current through both the emitter-base and collector-base junctions. The collector-emitter breakdown voltage test determines the condition of both junctions simultaneously.

Breakdown voltage is designated as BV_{CBO} (collector-base, emitter open), BV_{CES} (emitter shorted to base), or BV_{CEO} (collector-to-emitter, base open) on most data sheets. Breakdown is normally measured with the emitter (or base) open, but can also be measured with the emitter shorted to the base,

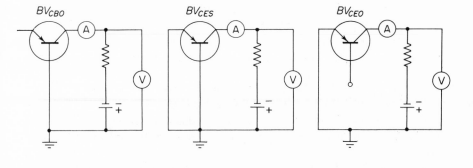

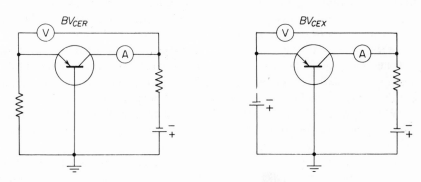

Fig. 3-2. Breakdown voltage test circuits.

connected to the base through a resistor, or with the emitter and base reverse biased.

Figure 3-2 shows the basic circuits for breakdown tests. The circuits shown are for *PNP* transistors. The same circuits can be used for *NPN* transistors when the voltage source polarity is reversed.

In all cases, the voltage source is adjusted for a given leakage current flow. Then the voltage is compared with a minimum specified voltage.

3-4. Gain Tests

The dynamic gain of a transistor is determined by the amount of change in output for a given change in input. Usually, transistors are tested for *current gain*. The change in output current for a given change in input current is measured, without changing the output voltage.

When a transistor is connected in a common base circuit, the collector forms the output circuit, while the emitter forms the input circuit. Common base current gain is known as Alpha, indicated by the Greek letter (α).

Most data sheets now specify gain with the transistor connected in a common-emitter circuit, rather than a common base. In the common-emitter circuit, the base is the input, and the collector is the output. Current gain for a common-emitter circuit is known as Beta, indicated by the Greek letter (β).

In addition to Alpha and Beta, present-day data sheets use several other terms to specify gain. The term "forward current transfer ratio" and the letters h_{fe} are the most popular means of indicating current gain for transistors, even though some manufacturers use "collector-to-base current gain."

The h in the letters h_{fe} refers to the hybrid (mixture) of transistor equivalent model circuits. Transistor test circuits are structured along the same lines. The hybrid system is discussed in Chapter 13. In the hybrid system, the transistor and the test or operating circuit are considered as a "black box" with an input and an output, rather than individual components.

When lower case letters h_{fe} (or sometimes H_{fe}) are used in transistor specifications, this indicates that the current gain is measured by noting the change in collector alternating current, for a given change in base alternating current. This is also known as "a-c Beta" or "dynamic Beta."

When capital letters H_{FE} are used in transistor specifications, the current gain is measured by noting the collector direct current for a given base direct current. This is also known as "d-c Beta."

Direct-current gain measurements apply under a wider range of conditions, and are easier to make. Alternating current gain measurements require more elaborate test circuits, and the test results will vary with the frequency of the alternating current used for the test. However, a-c measurements are more realistic, since transistors are usually used with a-c signals.

There are a number of circuits for the transistor gain test, and a number of test procedures. Likewise, there are many commercial transistor testers, as well as adapters which permit transistors to be tested with oscilloscopes. Some testers even permit transistors to be tested while still connected in the circuit. It would be impractical to cover the use and operation of all such testers and oscilloscope adapters in this book. Also, detailed operating instructions are provided with the testers. These instructions should be followed in all cases. Instead of attempting to duplicate the operating instructions, the following paragraphs describe the *operating principles* of the tests.

Basic Transistor Gain Tests

Figure 3-3 shows the basic circuits for Alpha measurements of *PNP* and *NPN* transistors. Under static conditions, both the emitter current I_e and collector current I_c are measured. Then the emitter current I_e is changed

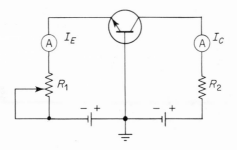

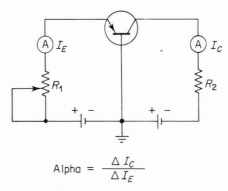

$$\text{Alpha} = \frac{\Delta I_C}{\Delta I_E}$$

Fig. 3-3. Basic d-c Alpha test circuits.

a given amount by varying the resistance of R_1, or by changing the emitter-base source voltage. The collector voltage must remain the same.

The *difference* in collector current I_c is noted, and the value of Alpha is calculated using the equation shown. For example, assume that the emitter current I_e is changed 4 milliamperes, and that this results in a change of 1 milliampere in collector current I_c. This would mean a current gain of 0.25.

Figure 3-4 shows the basic circuits for Beta measurements of *PNP* and *NPN* transistors. Under static conditions, both the base current I_b and the collector current I_c are measured. Then, without changing the collector voltage, the base current I_b is changed by a given amount, and the difference in collector current I_c is noted.

For example, assume that when the circuit is first connected the base current I_b is 7 milliamperes, and the collector current I_c is 43 milliamperes. When the base current I_b is increased to 10 milliamperes (a 3-milliampere increase), the collector current I_c increases to 70 (a 27-milliampere increase). This represents a 27-milliampere increase in collector current I_c for a 3-milliampere increase in base current I_b, or a current gain of 9.

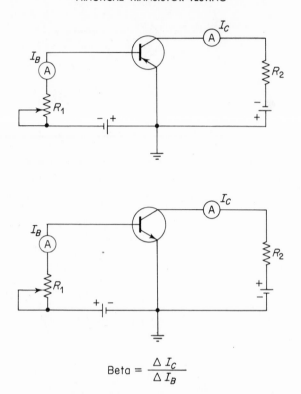

$$Beta = \frac{\Delta I_C}{\Delta I_B}$$

Fig. 3-4. Basic d-c Beta test circuits.

NOTE

Certain precautions should be observed if transistors are to be tested using *noncommercial* test circuits. The most important of these are as follows:

1. The collector and emitter (or base) load resistances (represented by R_1 and R_2 in Figs. 3-3 and 3-4) should be of such value that the maximum current limitations of the transistor are not exceeded (as discussed in Sec. 3-2). In the case of power transistors, the wattage rating of the load resistance should be large enough to dissipate the heat.

2. Where there is a large collector leakage current, this must be accounted for in test conditions.

3. The effect of meters used in the test circuits must also be taken into account.

Basic Transistor A-C Gain Tests

There are several types of circuits used for the a-c or dynamic test of transistors. Some of the commercial transistor testers use the same basic circuits shown in Figs. 3-3 and 3-4, except that an alternating current signal is introduced into the input, and the gain is measured at the output. Usually, these testers provide a 60 Hz- or 1000-Hz a-c signal for injection into the input. Where it is desirable to test transistors at higher frequencies, some tester circuits permit an external high frequency signal to be injected.

One of the most common methods used in shop-type (nonlaboratory) transistor testers for a-c gain measurement is the *feedback* circuit. A typical feedback circuit is shown in Fig. 3-5. In this circuit, the transistor under test is inserted into an audio oscillator configuration. The amount of feedback is adjusted by means of a calibrated control until the circuit begins to oscillate. Oscillation is indicated by a tone on the loudspeaker. The transistor current gain at the oscillation starting point can be read directly from the dial calibration.

When power is first applied to the circuit, the collector current starts to increase. This increase in collector current is fed back to the transformer T_1 primary through variable resistor R_1 (the shaft of which is coupled to the calibrated dial). Any change in the T_1 primary current causes a corresponding change in the T_1 secondary current. This change is applied to the base of the transistor under test, and causes a change in base current. Any change in base current causes a greater change in collector current which, in turn, is fed back to the transformer and base to produce further change. The process

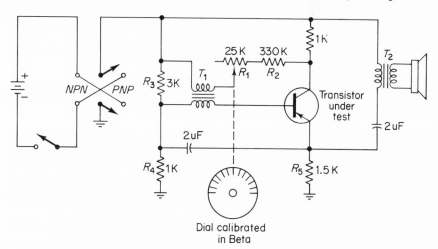

Fig. 3-5. Feedback type gain (Beta) test circuit.

continues until the circuit starts to oscillate at an audio frequency that is heard on the loudspeaker.

The setting of the feedback control, and the transistor gain, determine the point at which oscillation will start. For example, if a large feedback is required to produce oscillation, the transistor gain is low. If the circuit will oscillate with very little feedback, the transistor gain is high. Therefore, the dial connected to variable feedback resistor R_1 can be calibrated directly in terms of gain.

Laboratory Transistor A-C Gain Tests

The most practical means of measuring transistor gain in the laboratory is to display the transistor characteristics as curve traces on an oscilloscope. Since the oscilloscope screen can be calibrated in voltage and current, the transistor characteristics can be read off the screen directly. If a number of curves are made with an oscilloscope, they can be compared with the curves drawn on transistor data sheets.

There are a number of oscilloscopes (or oscilloscope adapters) manufactured specifically to display transistor curves. The Tektronix Transistor Curve-Tracer is a typical unit. Transistors under test are inserted into either a common-base or common-emitter test circuit. The transistor collector has a sweep voltage applied to it, while a step voltage or current is applied to either the base or emitter (whichever is ungrounded). Voltage (for the collector) sweeps between zero and a selectable value. Step voltages (for the emitter or N base) start at zero and build up to a value determined by the number of steps and value of each step as selected. Each sequence of steps, from zero to the maximum attained value, in conjunction with the sweep voltage on the collector, produces one family of characteristic curves.

Signals used for vertical and horizontal deflection on the oscilloscope screen are either current or voltage values selected from various points in the transistor test circuit. Thus, a selected vertical signal can be plotted against a selected horizontal signal to trace the desired transistor characteristic curve. A laboratory curve tracer contains circuits which permit almost any combination of collector, emitter, or base voltage and current to be displayed.

These circuits can be duplicated using a d-c oscilloscope for display. The most important set of curves are those that show collector current versus emitter current. Figure 3-6 illustrates the basic circuit required for such a display. For the circuit shown in Fig. 3-6, both the vertical and horizontal channels of the oscilloscope must be *voltage calibrated.* Usually, the horizontal channel of an oscilloscope is calibrated with respect to time. The horizontal and vertical channels must be identical, or nearly identical, to eliminate any phase difference. The horizontal zero reference point should be at the left (or right) of the oscilloscope screen rather than in the center.

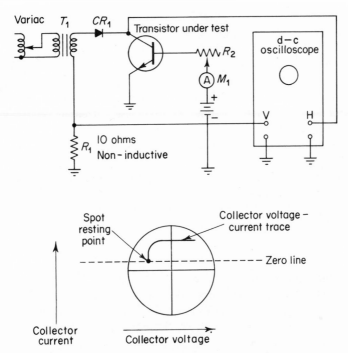

Fig. 3-6. Testing NPN transistors for collector current versus input (common emitter) current.

As shown in Fig. 3-6, the transistor is tested by applying a controlled d-c voltage to the collector. The collector voltage is developed by rectifying the transformer T_1 secondary voltage with diode CR_1, and can be adjusted to any desired value by the variac. When collector current flows on positive half cycles, the current flows through R_1. The voltage drop across R_1 is applied to the vertical channel, and causes the spot to move up and down. Therefore, vertical deflection is proportional to current. The vertical scale divisions can be converted directly to current when R_1 is made 1 ohm. With R_1 at a value of 10 ohms as shown in Fig. 3-6, the indicated voltage value must be divided by 10 to obtain current. For example, a 3-volt vertical deflection indicates a 0.3 ampere current.

The same voltage applied to the transistor collector is applied to the horizontal channel (which has been voltage calibrated) and causes the spot to move from left to right (for the *NPN* transistor shown). Therefore, horizontal position is proportional to voltage. The combination of the horizontal (voltage) deflection and vertical (current) deflection causes the spot to trace out the collector current—collector voltage characteristics of the transistor.

Usually, the change in collector current for a given change in emitter-base current (or Beta) is the desired characteristic for most transistors. This can be displayed by setting the emitter-base current to a given value and measuring the collector-current curve, with a given collector voltage. Then the emitter-base current is changed to another value, and the new collector current is displayed, without changing the collector voltage. Collector voltage is set by the variac. Emitter-base current is set by R_2, and measured on M_1.

On the commercial transistor curve tracers, the emitter-base current is applied in steps.

The test connection diagram of Fig. 3-6 is for an *NPN* transistor connected in a common emitter circuit. If a *PNP* transistor is to be tested, the polarity of rectifier CR_1, battery B, and meter M_1 must be reversed. Also, the horizontal zero reference point should be at the right of the screen rather than at the left.

The following procedure will display a *single curve*:

1. Connect the equipment as shown in Fig. 3-6.

2. Place the oscilloscope in operation. Voltage calibrate both the vertical and horizontal channels as necessary. The spot should be at the vertical center, and at the left (for *NPN*) of the horizontal center with no signal applied to either channel.

3. Switch off the internal recurrent sweep. Set sweep-selector and sync-selector to external. Leave the horizontal- and vertical-gain controls set at the (voltage) calibrate position. Set the vertical polarity switch so that the trace will deflect up from the center line as shown in Fig. 3-6.

4. Adjust the variac so that the voltage applied to the collector is the maximum rated value. This voltage can be read on the voltage-calibrated horizontal scale.

5. Adjust resistor R_2 for the desired emitter-base current as indicated on meter M_1.

6. Check the oscilloscope pattern against the transistor specifications. Compare the current-voltage values against the specified values. For example, assume that a collector current of 300 milliamperes should flow with 7 volts applied. This can be checked by measuring along the horizontal scale to the 7-volt point, then measuring from that point up the trace. The 7-volt (horizontal) point should intersect the trace at the 300-milliampere (3-volt) vertical point.

7. If desired, adjust resistor R_2 for another emitter-base current value as indicated on meter M_1. Then check the new collector current-voltage curve.

Interpreting Transistor Traces

Transistor curve traces can be used to determine a number of characteristics. For example, the traces shown in Fig. 3-7 are for a 2N338 transistor

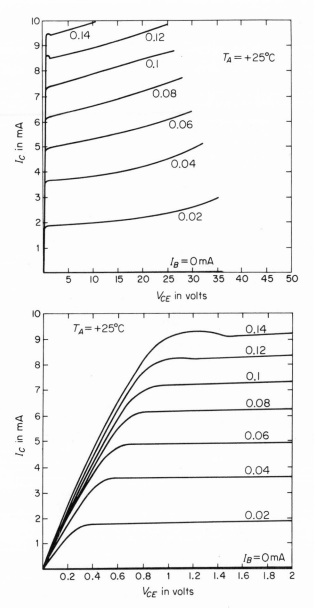

Fig. 3-7. Collector output characteristics for a 2N338 transistor at 25°C.

operating at a temperature of 25°C, while the curves of Fig. 3-8 are for the same transistor at 150°C. Assume that it is desired to find the transistor gain with 20 volts applied to the collector.

At an ambient temperature of 25°C, and an input current of 0.02 milliamperes, the collector current is approximately 2.2 mA. When the input

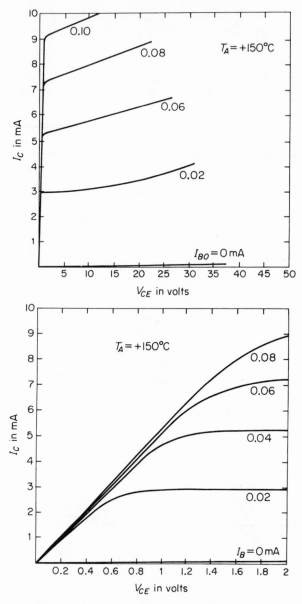

Fig. 3-8. Collector output characteristics of a 2N338 transistor at 150°C.

current is increased to 0.04 mA, the collector current increases to 4.2 mA. This indicates a 2-mA collector-current increase for a 0.02-mA increase in input current. Therefore, the gain is 100 (2/0.02 = 100).

The gain of a transistor will vary with temperature, frequency, and different levels of input.

For example, assuming the same collector voltage (20 volts) and ambient temperature (25°C), the collector current is approximately 8.5 mA when the input current is 0.1 mA. When the input current is increased to 0.12 mA, the collector current increases to 9.5 mA. This indicates a 1-mA collector-current increase for a 0.02-mA increase in input current. Therefore, the gain is 50 (1/0.02 = 50).

Assuming the same collector voltage (20 volts), but a different ambient temperature (150°C), the collector current is approximately 3.5 mA when the input current is 0.02 mA (see Fig. 3-8). When the input current is increased to 0.06 mA, the collector current increases to 6.5 mA. This indicates a 3-mA collector-current increase for a 0.04-mA increase in input current. Therefore, the gain is 75 (3/0.04 = 75).

Transistor curve traces can also be used to determine the value of the load resistor for basic transistor circuits. This is known as drawing a *load line*. The transistor curves shown in Fig. 3-9 are for a *PNP* transistor circuit such as shown in Fig. 3-10. Since the base is grounded, the emitter-base (or input) current is designated as emitter current, or I_E. As discussed in

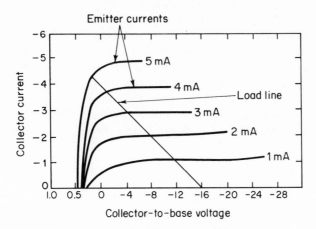

Fig. 3-9. Typical *PNP* transistor curves.

Fig. 3-10. Simplified representation of characteristics shown by *PNP* transistor curves.

Chapter 2, current gain for a grounded-base emitter circuit (or Alpha) is less than 1, but power amplification occurs because of the higher impedance in the output circuit. Since power is equal to I^2R, the higher output resistance versus the very low input resistance of the grounded base circuit produces signal power amplification.

Both the collector current (vertical axis) and collector-to-base voltage (horizontal axis) are shown on Fig. 3-9 as negative values, since the transistor is a *PNP*. If an *NPN* were used, positive values of collector current and voltage would be shown.

At audio frequencies, the load resistance R_L of Fig. 3-10 is represented as an actual resistor no matter what type of coupling (such as transformer) is used to the next stage. The actual load reflected back to the transistor from the next stage (or load such as a loudspeaker) represents pure resistance (which consumes power).

The load line of Fig. 3-9 can show the collector voltage and current swing for various input currents. When the input signal swings in a direction (positive, in this case) to increase the forward bias on the emitter-base circuit, the collector current increases, increasing the drop across the load resistor, and reducing the collector voltage. For example, with a 1-mA input, the collector voltage is -12 volts, and the collector current is -1 mA. When the input current is increased to 2 mA, the collector voltage drops to -8 volts, and the collector current is -2 mA. If the positive swing of the input signal is great enough to forward bias the input to a point where the input current is about 4.5 mA, the collector voltage will drop to zero (theoretically), and the collector current will be -4 mA. Therefore, the collector current swing is from zero to -4 mA, while the collector voltage swing is from zero to -16 volts.

The load resistance represented by the load line of Fig. 3-9 can be calculated as follows:

$$\text{Load resistance} = \frac{\begin{array}{cc}\text{Maximum} & \text{Minimum} \\ \text{collector} & - \ \text{collector} \\ \text{voltage} & \text{voltage}\end{array}}{\begin{array}{cc}\text{Maximum} & \text{Minimum} \\ \text{collector} & - \ \text{collector} \\ \text{current} & \text{current}\end{array}} = \frac{16 - 0}{0.004 - 0} \text{ or } 4000$$

The power output (or signal power) of the basic circuit shown in Fig. 3-10 can also be determined by the load line of Fig. 3-9 as follows:

$$\text{Power output} = \frac{\left[\begin{array}{cc}\text{Maximum} & \text{Minimum} \\ \text{collector} & - \ \text{collector} \\ \text{voltage} & \text{voltage}\end{array}\right] \times \left[\begin{array}{cc}\text{Maximum} & \text{Minimum} \\ \text{collector} & - \ \text{collector} \\ \text{current} & \text{current}\end{array}\right]}{8}$$

or

$$\frac{(16 - 0) \times (0.004 - 0)}{8} = 0.08 \text{ watt}$$

The load line information also establishes the quiescent *operating point* (also known as the *center point*, or *zero signal point*) of the transistor. If a resistor were used as the transistor circuit load, the voltage drop across the resistor would vary from zero to 16 volts. If the input circuit were forward biased so that 2 mA of emitter current would flow when there was no signal, the collector voltage would be -8 volts. This establishes the operating point as -8 volts. A positive swing in input signal will drive the collector voltage toward zero, while a negative input signal moves the collector voltage toward the source voltage of -16 volts.

The amount of swing on either side of the operating point is determined by the input signal strength and the forward bias on the input circuit. Assume that it is desired to operate the transistor so that the collector voltage swings the full 16 volts (from zero to -16) when a full input signal is applied. (This is approximately Class A operation. In true Class A operation, the collector voltage swing would stop short of -16 volts, and there would always be some collector current flowing.)

A positive input signal will add to the forward bias, causing more collector current to flow, and move the collector voltage toward zero. (This is actually a positive voltage swing since the voltage becomes less negative.) A negative input signal will oppose the forward bias, reducing collector current flow, and move the collector voltage toward -16 volts. Assume that the input signal swings from a maximum of -1 volt to $+1$ volt. Then a forward bias of 1 volt would be required. On positive input signals, the two voltages (input and forward bias) would combine to produce an input of 2 volts. On negative input signals, the two voltages would oppose each other to produce an input of zero.

Assuming a 1-volt forward bias, a 500-ohm input resistor would be required, since a 2-mA input (emitter) current is required to produce a -8 collector voltage ($500 = 1/0.002$).

Therefore, under no-signal conditions (or at the zero point of the input signal), the 1-volt forward bias on the input current produces a 2-mA current flow through the 500-ohm input resistor. The 2-mA input (emitter) current causes a -2-mA collector current to flow through the 4000-ohm load resistor, resulting in an 8-volt drop across the load. The collector voltage is then -8 volts (at the operating point).

Under the maximum negative swing of the input voltage, the -1-volt input signal opposes the $+1$-volt forward bias, and (theoretically) no current flows in the input circuit. This causes the collector current to stop flowing (theoretically), and the collector voltage swings to the -16-volt point.

Under the maximum positive swing of the input voltage, the $+1$-volt input signal adds to the $+1$-volt forward bias, and produces a 4-mA (approximate) current flow through the 500-ohm input resistor. This causes a -4-mA (approximate) collector current to flow through the 4000-ohm load resistor, resulting in a 16-volt drop across the load. The collector voltage is then zero.

If a *transformer* were used as the transistor circuit load, instead of a resistor, the operating voltage would be equal to the source voltage. In a transformer-coupled circuit, the actual load is that which is applied to the secondary of the transformer, such as a loudspeaker voice-coil. The function, in terms of collector current swing, occurs because of the inductance through which the collector current must flow. If the negative swing of the input signal is sufficient to reduce the collector current to a small value, the sudden change in current through the inductance induces a voltage across this inductance, which adds to the collector supply voltage. The effect is to increase the instantaneous collector voltage to a value considerably higher than the operating voltage (or source voltage).

So long as the *average d-c collector current* does not change when the input signal is applied, the same dynamic characteristics hold true for both transformer-coupled and pure resistive loads. However, the transformer-coupled circuit can operate with one-half the collector-source voltage. In the examples of Figs. 3-9 and 3-10, a transformer-coupled circuit would require -8 volts, while a circuit with a pure resistive load would require -16 volts. This demonstrates the efficiency of transformer-coupled circuits.

No matter what circuit configuration is used, or what source voltage is applied, the maximum voltage/current combination must never exceed the maximum power rating for the transistor.

3-5. Switching Tests

Transistors to be used in pulse or digital applications must be tested for switching characteristics. For example, when a pulse is applied to the input of a transistor, there is a measurable time delay before the pulse starts to appear at the output. Likewise, after the pulse is removed, there is additional time delay before the transistor output returns to its normal level. These "switching times" or "turn on" and "turn off" times are usually in the order of a few microseconds for high-speed pulse transistors.

The switching characteristics of transistors designed for computer or industrial work are listed on the data sheets. Each manufacturer lists his own set of specifications. However, there are four terms (rise time, fall time, delay time, and storage time) common to most data sheets for transistors used in pulse work. These switching characteristics are of particular importance where the pulse durations are short. For example, assume that the "turn on" time of a transistor is 10 microseconds, and that a 5-microsecond pulse were applied to the transistor input. There would be no output pulse, or the pulse would be drastically distorted, under these conditions.

Pulse and Squarewave Definitions

The following terms are commonly used in describing transistor switching characteristics. The terms are illustrated in Fig. 3-11. The input pulse repre-

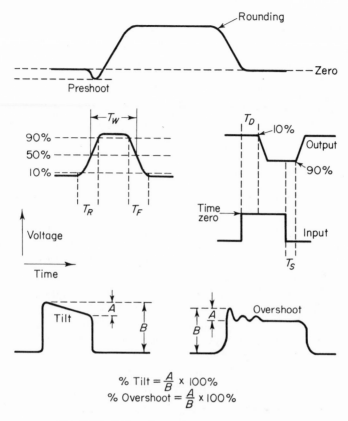

$$\% \text{ Tilt} = \frac{A}{B} \times 100\%$$
$$\% \text{ Overshoot} = \frac{A}{B} \times 100\%$$

Fig. 3-11. Basic pulse and squarewave definitions.

sents an ideal input waveform for comparison purposes. The other waveforms in Fig. 3-11 represent typical output waveforms in order to show the relationships. The terms are defined as follows:

Rise Time T_R: the time interval during which the amplitude of the output voltage changes from 10 per cent to 90 per cent of the rising portion of the pulse.

Fall Time T_F: the time interval during which the amplitude of the output voltage changes from 90 per cent to 10 per cent of the falling portion of the waveform.

Time Delay T_D: the time interval between the beginning of the input pulse (time zero), and the time when the rising portion of the output pulse attains an arbitrary amplitude of 10 per cent above the baseline.

Storage Time T_S: the time interval between the end of the input pulse (trailing edge), and the time when the falling portion of the output pulse drops to an arbitrary amplitude of 90 per cent from the baseline.

Pulse Width (or Pulse Duration) T_W: the time duration of the pulse measured between two 50-per cent amplitude levels of the rising and falling portions of the waveform.

Tilt: a measure of the tilt of the full-amplitude flat-top portion of a pulse. The tilt measurement is usually expressed as a percentage of the amplitude of the rising portion of the pulse.

Overshoot: a measure of the overshoot occurring generally above the 100-per cent amplitude level. This measurement is also expressed as a percentage of the pulse rise.

These definitions are for guide purposes only. When pulses are very irregular (such as excessive tilt, overshoot, etc.), the definitions may become ambiguous.

Testing Transistors for Switching Time

An oscilloscope having wide-frequency response, good transient characteristics, and a dual trace can be used to check the high-speed switching characteristics of transistors used in pulse or computer work. The oscilloscope vertical channel must be voltage calibrated in the normal manner, while the horizontal channel should be time calibrated (rather than sweep frequency calibrated).

As shown in Fig. 3-12, the transistor is tested by applying a pulse to the base of the transistor under test. This same pulse is applied to one of the oscilloscope vertical inputs. The transistor collector output is applied to the other oscilloscope vertical input (inverted 180° by the common-emitter circuit). The two pulses are then compared as to rise time, fall time, delay time, storage time, etc. The transistor output-pulse characteristics can then be compared with the transistor specifications.

1. Connect the equipment as shown in Fig. 3-12.

2. Place the oscilloscope in operation. Switch on the oscilloscope internal recurrent sweep. Set the sweep-selector and sync-selector to internal.

3. Set the pulse generator to produce a 2.2-volt, 3-microsecond positive

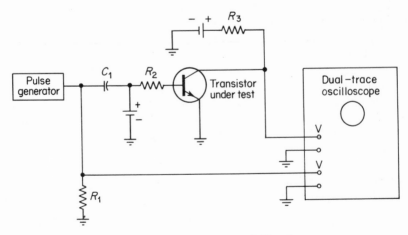

Fig. 3-12. Testing transistor switching time.

pulse. These values are taken (arbitrarily) from the data sheet of a General Electric 2N338 transistor, but are typical for many computer transistors. Always use the values specified in the data sheet.

4. Adjust the collector and base supply voltages to the values specified in the transistor manufacturer's data.

5. The oscilloscope pattern should appear as shown in Fig. 3-12, with both the transistor output pulse and input pulse displayed.

6. Measure the rise time, fall time, delay time, and storage time along the horizontal (time-calibrated) axis of the oscilloscope display.

Rule of Thumb for Switching Tests

Since rise-time (and fall-time) measurements are of special importance in switching tests, the relationship between oscilloscope rise time, and the rise time of the transistor (or other device such as a diode) must be taken into account. Obviously, the accuracy of rise-time measurements can be no greater than the rise time of the oscilloscope. Also, if the device is tested by means of an external pulse from a pulse generator, the rise time of the pulse generator must also be taken into account.

For example, if an oscilloscope with a 20-nanosecond rise time is used to measure the rise time of a 15-nanosecond transistor, the measurement would be hopelessly inaccurate. Likewise, if a 20-nanosecond pulse generator and a 15-nanosecond oscilloscope were used to measure the rise time of a device, the fastest rise time for accurate measurement would be something greater than 20 nanoseconds.

There are two basic rules of thumb that can be applied to rise-time measurements.

The first method is known as the "root of the sum of the squares." It involves finding the square of all the rise times associated with the test, adding these squares together, and the finding the square root of this sum. For example, using the 20-nanosecond pulse generator and the 15-nano-second oscilloscope, the calculation would be as follows:

$$20 \times 20 = 400; \ 15 \times 15 = 225; \ 400 + 225 = 625;$$

$$\sqrt{625} = 25 \text{ (nanoseconds)}$$

One major drawback to this rule is that the coaxial cables required to interconnect the test equipment are subject to "skin effect." As frequency increases, the signals tend to travel on the outside or skin of the conductor. This decreases conductor area, and increases resistance. In turn, this increases cable loss. The losses of cables do not add properly when applied to the root-sum-squares method, except as an approximation.

The second rule or method states that if the equipment or signal being measured has a rise time *10 times* slower than the test equipment, the error is 1 per cent. This is small and can be considered as negligible. If the equipment being measured has a rise time *3 times* slower than the test equipment, the error is slightly less than 6 per cent.

CHAPTER 4

Practical Diode Testing

This chapter describes diode characteristics and test procedures from the practical, technician-level standpoint. That is, the data in this chapter will permit the technician to test all of the *important* diode characteristics, and to understand the basis for such tests. Both power-rectifier diodes and small-signal diodes are covered, as are Zener and tunnel diodes. Semiconductor-controlled rectifier diodes are covered in Chapter 9, while voltage-variable diodes are covered in Chapter 11.

4-1. Basic Diode Tests

As in the case of transistors, all types of diodes are subjected to a variety of tests during manufacture. It is neither practical nor necessary to duplicate all of these tests in the field. For power-rectifier diodes and small-signal diodes, there are three basic tests required. First, any diode must have the ability to pass current in one direction (forward current), and prevent (or limit) current flow (reverse current) in the opposite direction. Second, for a given reverse voltage, the reverse current should not exceed a given value. Third, for a given forward current, the voltage drop across the diode should not exceed a given value.

If the diode is to be used in pulse or digital work, the switching time must also be tested.

In addition to the basic tests, a Zener diode must also be tested for the correct *Zener voltage* point. Likewise, a tunnel diode must be tested for its *negative resistance* characteristics.

4-2. Diode Continuity Tests

The elementary purpose of a diode (both power-rectifier and small-signal) is to prevent current flow in one direction, while passing current in the opposite direction. The simplest test of a diode is to measure current flow in the forward direction with a given voltage, then reverse the voltage and measure current flow (if any). If the diode will prevent current flow in the reverse direction, but will pass current in the forward direction, the diode will meet most *basic* circuit requirements. If there is some current flow in the reverse direction (known as *leakage current*), it is still possible that the diode will operate properly in noncritical circuits.

A simple resistance measurement, or continuity check, can often be used to test a diode's ability to pass current in one direction only. A simple ohmmeter can be used to measure the forward and reverse resistance of a diode. Figure 4-1 shows the basic circuit. A good diode will show high resistance in the reverse direction, and low resistance in the forward direction. If the resistance is high in both directions, the diode is probably open. A low resistance in both directions usually indicates a shorted diode.

It is possible for a defective diode to show a difference in forward and reverse resistance. The important factor in making a diode resistance test is the *ratio* of forward-to-reverse resistance (often known as *front-to-back* ratio). The actual ratio will depend upon the type of diode. However, as a rule of thumb, a small signal diode will have a ratio of several hundred to one, while a power rectifier can operate satisfactorily with a ratio of 10-to-1.

Diodes used in power circuits are usually not required to operate at high frequencies. Such diodes may be tested effectively with direct current or low-frequency alternating current. Diodes used in other circuits, even audio

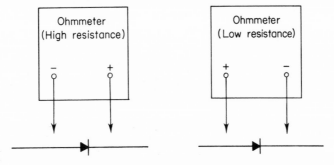

Fig. 4-1. Basic ohmmeter test of diodes for front-to-back ratio.

equipment, must be capable of operation at higher frequencies, and should be so tested.

Commercial Diode Testers

Most simple commercial diode testers operate on the *continuity test* principle. This is similar to testing a diode by measuring resistance, except that the actual resistance value is of no concern.

Figure 4-2 shows the basic circuit of a simple commercial diode tester using the continuity principle. The condition of the diode under test is indicated by lamps. With an open diode, or no connection across the test terminals, lamp L_1 lights due to the current flow through CR_1, R_1, and R_2. The voltage drop across R_2 (positive at the top end) is too low to light lamp L_3. Since CR_2 is reverse biased, lamp L_2 does not light.

If a good diode is connected across the test terminals with the polarity as shown in Fig. 4-2, lamp L_1 is shorted out and does not light. Also, lamp L_2 does not light because CR_2 is reverse-biased by the voltage developed by the diode under test.

Capacitor C_1 charges, permitting lamp L_3 to light, thus indicating that the diode is good.

A shorted diode will also short out lamp L_1, and will allow both half-cycles of the alternating current to be applied across R_2. CR_2 will conduct on the negative half-cycles, causing lamp L_2 to light. Alternating current is also applied across R_4, R_5, L_3, C_1, and C_2. The capacitors C_1 and C_2 appear as a short across R_5 and L_3. Therefore, lamp L_3 does not light. This leaves only lamp L_2 to light and indicate a short.

If a good diode is connected into the test terminals with the polarity

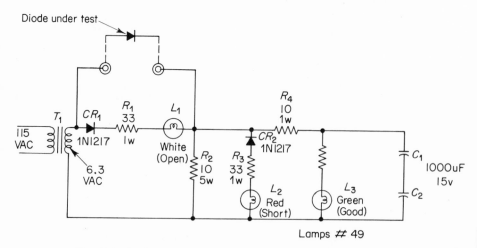

Fig. 4-2. Continuity-type diode test circuit (*courtesy General Electric*).

reversed, all lamps will light. Under these conditions, CR_1 conducts on positive half-cycles, causing lamp L_1 to light. The test diode conducts on negative half-cycles, and develops a d-c voltage across R_2 (with the top end negative). CR_2 then conducts and L_2 lights. Capacitor C_2 charges and permits L_3 to light.

4-3. Reverse Leakage Tests

Reverse leakage is the current flow through a diode when a reverse voltage (anode negative) is applied. The basic circuit for measurement of reverse leakage is shown in Fig. 4-3. Similar circuits are incorporated in some advanced diode testers, or can be duplicated with the simple test equipment shown.

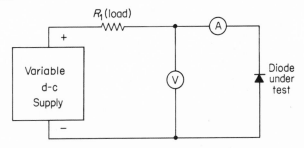

Fig. 4-3. D-C reverse leakage test for diodes.

As shown in Fig. 4-3, the diode under test is connected to a variable direct current source in the reverse-bias condition (anode negative). The variable source is adjusted until the desired voltage is applied to the diode as indicated by the voltmeter. Then the current (if any) through the diode is measured by the current meter. This is the reverse (or leakage) current. Usually, excessive leakage current is undesired, but the limits should be determined by reference to the manufacturer's data sheet.

4-4. Forward Voltage Drop Tests

Forward voltage drop is the voltage that appears across the diode when a given forward current is being passed. The basic circuit for measurement of forward voltage is shown in Fig. 4-4. Similar circuits are incorporated in some advanced diode testers, or can be duplicated with simple test equipment as shown.

As shown in Fig. 4-4, the diode under test is connected to a variable direct current source in the forward-bias condition (anode positive). The variable source is adjusted until the desired amount of current is passing through the diode as indicated by the current meter. Then the voltage drop

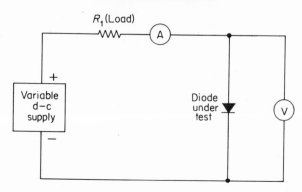

Fig. 4-4. D-C forward voltage drop test for diodes.

across the diode is measured by the voltmeter. This is the forward-voltage drop. Usually, a large forward-voltage drop is not desired. The maximum limits should be determined by reference to the manufacturer's data sheet.

4-5. Dynamic Diode Tests

The circuits and methods discussed in previous sections of this chapter provide a *static* test of diodes. That is, the diode is subjected to a constant direct current when the leakage and voltage drop are measured. Diodes rarely operate this way in circuits. Instead, diodes are operated with alternating current which tends to heat the diode junctions, and change the characteristics. It is more realistic to test a diode under *dynamic* conditions.

Checking Power Rectifier Diodes

Power rectifier diodes can be subjected to a dynamic test, using a direct current oscilloscope to display and measure the current-voltage characteristics. To do so, both the vertical and horizontal channels of the oscilloscope must be voltage calibrated. Usually, the horizontal channel of most oscilloscopes is time-calibrated. However, the horizontal channel can be voltage calibrated, using the same procedures as for the vertical channel. (Such procedures are described in the oscilloscope instruction manual.) Also, the horizontal and vertical channels must be identical, or nearly identical, to eliminate any phase difference.

As shown in Fig. 4-5, the power-rectifier diode is tested by applying a controlled a-c voltage across the anode and cathode, through resistor R_1. The a-c voltage (set to the maximum-rated peak-inverse voltage, or *PIV*, of the diode) alternately biases the anode positive and negative, causing both forward and reverse current to flow through R_1. The voltage drop across R_1 is applied to the vertical channel, and causes the screen spot to move up

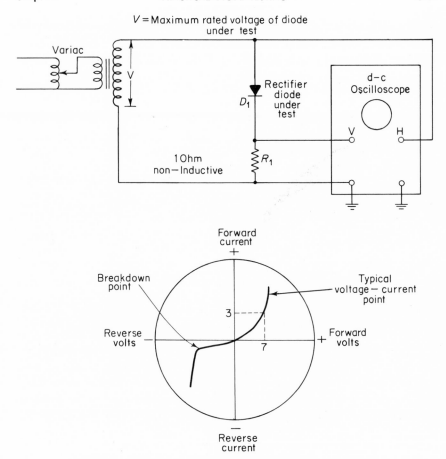

Fig. 4-5. Testing diode rectifier voltage-current characteristics.

and down. Vertical deflection is proportional to current. The vertical scale divisions can be converted directly to current when R_1 is made 1 ohm. For example, a 3-volt vertical deflection indicates a 3-ampere current.

The same voltage applied across the diode is applied to the horizontal channel (which has been voltage-calibrated), and causes the spot to move right or left. Horizontal deflection is proportional to voltage. The combination of the horizontal (voltage) deflection and vertical (current) deflection causes the spot to trace out the complete current-voltage characteristics.

The procedure is as follows:

1. Connect the equipment as shown in Fig. 4-5.

2. Place the oscilloscope in operation. Voltage calibrate both the vertical and horizontal channels as necessary. The spot should be at the vertical and horizontal center with no signal applied to either channel.

3. Switch off the internal recurrent sweep. Set sweep-selector and sync-selector to external. Leave the horizontal- and vertical-gain controls set at the (voltage) calibrate position.

4. Adjust the variac so that the voltage applied across the power rectifier D_1 under test is the maximum rated value.

5. Check the oscilloscope pattern against the typical curves of Fig. 4-5 and/or against the diode specifications. The curve of Fig. 4-5 is a typical response pattern. That is, the forward current increases as forward voltage increases. Reverse current increases only slightly as reverse voltage is applied, unless the breakdown or "avalanche" point is reached. In conventional (non-Zener) diodes, it is desirable, if not mandatory, to operate considerably below the breakdown point.

6. Compare the current-voltage values against the values specified in the diode data sheet. For example, assume that a current of 3 amperes should flow with 7 volts applied. This can be checked by measuring along the horizontal scale to the 7-volt point, then measuring from that point up (or down) to the trace. The 7-volt (horizontal) point should intersect the trace at the 3-ampere (vertical) point.

Checking Small Signal Diodes

The procedures for checking the current-voltage characteristics of a small signal diode are the same as for power rectifier diodes. However, there is one major difference. In a small signal diode, the ratio of forward voltage to reverse voltage is usually quite large. A forward voltage of the same amplitude as the rated reverse voltage will probably damage the diode. On the other hand, if the voltage is lowered for both forward and reverse directions, this will not provide a realistic value in the reverse direction.

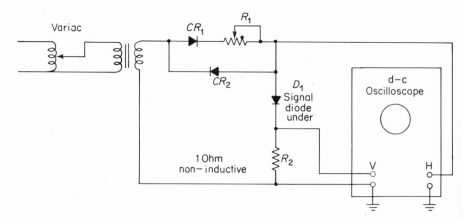

Fig. 4-6. Testing signal diode voltage-current characteristics.

Under ideal circumstances, a small signal diode should be tested with a low-value forward voltage, and a high-value reverse voltage. This can be accomplished using a circuit as shown in Fig. 4-6. It will be seen that the circuit of Fig. 4-6 is essentially the same as that of Fig. 4-5 (for power rectifier diodes), except that diodes CR_1 and CR_2 are included to conduct on alternate half-cycles of the voltage across transformer T_1. Rectifiers CR_1 and CR_2 are chosen for a linear amount of conduction near zero.

The variac is adjusted for maximum rated reverse voltage across the diode under test D_1, as applied through CR_2 when upper secondary terminal of T_1 goes negative. This applies the full reverse voltage.

Resistor R_1 is adjusted for maximum-rated forward voltage across the diode D_1, as applied through CR_1 when the upper secondary terminal of T_1 goes positive. This applies a forward voltage, limited by R_1.

With resistor R_1 properly adjusted, perform the current-voltage check as described for Power Rectifier Diodes.

4-6. Switching (Recovery Time) Tests

Diodes to be used in pulse or digital work must be tested for switching characteristics. The most important characteristic is *recovery time*. When a reverse-bias pulse is applied to a diode, there is a measurable time delay before the reverse current reaches its steady-state value. This delay period is listed as the recovery time (or some similar term) on the diode data sheet.

The duration of recovery time sets the minimum width for pulses with which the diode can be used. For example, if a 5-microsecond reverse-voltage pulse is applied to a diode with a 10-microsecond recovery time, the pulse will be distorted.

An oscilloscope having wide frequency response and good transient characteristics can be used to check the high speed switch and recovery time of diodes. The oscilloscope vertical channel must be voltage-calibrated in the normal manner, while the horizontal channel should be time-calibrated (rather then sweep-frequency calibrated).

As shown in Fig. 4-7, the diode is tested by applying a forward-biased current from the supply, adjusted by R_2 and measured by M_1. A negative squarewave is developed across R_3. This squarewave switches the diode voltage rapidly to a high negative value (reverse voltage). However, the diode does not cut off immediately. Instead, a steep transient is developed by the high momentary current flow. The reverse current falls to its steady-state value when the carriers are removed from the junction. This produces the waveform shown in Fig. 4-7.

Both forward and reverse currents are passed through resistor R_3. The voltage drop across R_3 is applied through emitter-follower Q_1 to the oscilloscope vertical channel. The coaxial cable provides some delay so that the

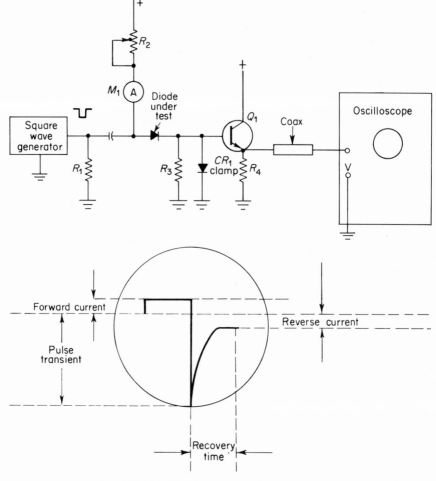

Fig. 4-7. Switching (recovery time) tests for diodes.

complete waveform will be displayed. CR_1 functions as a clamping diode to keep the R_4 voltage at a level safe for the oscilloscope.

The time interval between the negative peak and the point at which the reverse current has reached its low steady-state value, is the diode recovery time.

The procedure is as follows:

1. Connect the equipment as shown in Fig. 4-7.
2. Place the oscilloscope in operation.
3. Switch on the internal recurrent-sweep. Set sweep-selector and sync-selector to internal.
4. Set the squarewave generator to a repetition rate of 100 kHz, or as specified in the diode manufacturer's data sheet.

5. Set R_1 for the specified forward test current, as measured on M_1.

6. Increase the squarewave generator-output level (amplitude) until a pattern appears.

7. If necessary, readjust the sweep and sync controls until a single sweep is shown.

8. Measure the recovery time along the horizontal (time-calibrated) axis.

4-7. Zener Diode Tests

The test of a Zener diode is similar to that of a power rectifier or small-signal diode. The forward-voltage drop-test for a Zener is identical to that for a conventional diode, as described in Sec. 4-4. A reverse leakage test is usually not required, since a Zener will go into the avalanche condition when sufficient reverse voltage is applied. In place of a reverse leakage test, a Zener diode should be tested to determine the point at which avalanche occurs (establishing the Zener voltage across the diode). This can be done with a basic circuit, or with an oscilloscope. It is also common practice to test a Zener diode for its impedance. This is because the regulating ability of a Zener diode is related directly to its impedance. A Zener diode is similar to a capacitor in this respect; as the reactance decreases so does the change in voltage across the terminals.

Zener Voltage Tests

When sufficient reverse voltage is applied to a Zener diode, the avalanche condition will occur, and heavy reverse current will flow. If the external reverse voltage is increased, additional current will flow and drop the voltage to a fixed level. This is known as the Zener voltage.

The basic circuit for measurement of Zener voltage is shown in Fig. 4-8. The diode under test is connected to a variable direct-current source in the reverse-bias condition (anode negative). (This is the configuration in which

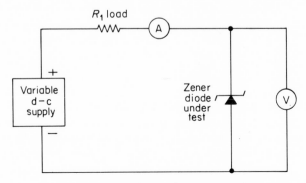

Fig. 4-8. Basic Zener voltage test circuit.

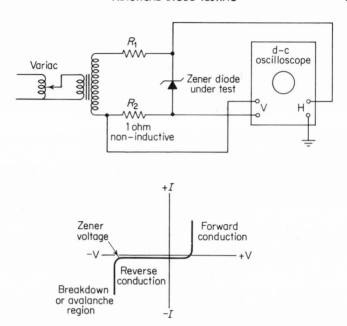

Fig. 4-9. Testing Zener diode voltage-current characteristics
(*courtesy International Rectifier*).

the Zener is normally used.) The variable source is adjusted until the Zener voltage is reached, and a large current is indicated through the current meter. Zener voltage can then be measured on the voltmeter.

The Zener characteristics can also be displayed on an oscilloscope. The procedure and circuit are similar to that for dynamic test of conventional diodes. In Fig. 4-9, the Zener diode is tested by applying a controlled a-c voltage across the anode and cathode, through resistors R_1 and R_2. The a-c voltage (set to some value above the Zener voltage) alternately biases the anode positive and negative, causing both forward and reverse current to flow through R_1 and R_2.

The voltage drop across R_2 is applied to the vertical channel, and causes the screen spot to move up and down. Vertical deflection is proportional to current. The vertical scale divisions can be converted directly to current when R_1 is made 1 ohm. For example, a 3-volt vertical deflection indicates a 3-ampere current.

The same voltage applied across the diode (taken from the junction of R_1 and the diode under test) is applied to the horizontal channel (which has been voltage-calibrated), and causes the spot to move right or left. Horizontal deflection is proportional to voltage. The combination of the horizontal (voltage) deflection and vertical (current) deflection causes the spot to trace out the complete current-voltage characteristics.

The procedure is as follows:

1. Connect the equipment as shown in Fig. 4-9.

2. Place the oscilloscope in operation. Voltage calibrate both the vertical and horizontal channels as necessary. The spot should be at the vertical and horizontal center with no signal applied to either channel.

3. Switch off the internal recurrent sweep. Set sweep-selector and sync-selector to external. Leave the horizontal- and vertical-gain controls set at the (voltage) calibrate position.

4. Adjust the variac so that the voltage applied across the Zener diode and the resistors R_1 and R_2 in series is greater than the rated Zener voltage.

5. Check the oscilloscope pattern against the typical curves of Fig. 4-9 and/or against the diode specifications. The curve of Fig. 4-9 is a typical response pattern. That is, the forward current increases as forward voltage increases. Reverse (or leakage) current increases only slightly as reverse voltage is applied, until the breakdown voltage is reached. Then an avalanche of current occurs.

6. Compare the current-voltage values against the values specified in the Zener diode data sheet. For example, assume that avalanche current should occur when the reverse voltage reaches 7.5 volts. This can be checked by measuring along the horizontal scale up to the 7.5-volt point, and noting that the current increases rapidly.

Zener Impedance Tests

As discussed, the regulating ability of a Zener diode is directly related to the diode's impedance. Likewise, Zener impedance varies with junction current and diode size. Therefore, to test a Zener diode properly for impedance, measurements must be made with a specific set of conditions. This can be accomplished using the circuit of Fig. 4-10.

With such a circuit, the diode direct current is set to approximately 20 per cent of the Zener maximum current by adjustment of R_1. The Zener direct

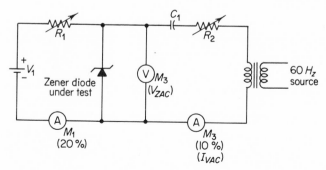

Fig. 4-10. Testing Zener diode impedance (*courtesy International Rectifier*).

current is indicated by meter M_1. Alternating current is also applied to the Zener, and is adjusted by R_2 to approximately 10 per cent of the maximum current rating of the diode. The Zener alternating current is indicated by meter M_3.

When these conditions have been met, the a-c voltage developed across the Zener junction can be read on meter M_2. When Zener a-c voltage (V_{zac}) and Zener alternating current (I_{zac}) are known, the impedance (Z_z) may be calculated using the equation:

$$Z_z = \frac{V_{zac}}{I_{zac}}$$

4-8. Tunnel Diode Tests

A tunnel diode must be tested for its *negative resistance* characteristics. The most effective test of a tunnel diode is to display the entire forward voltage-current characteristics on an oscilloscope. Thus, the valley and peak voltages, as well as the valley and peak currents, can be measured simultaneously.

A d-c oscilloscope is required. Both the vertical and horizontal channels must be voltage-calibrated. Also, the horizontal and vertical channels must be identical, or nearly identical, to eliminate any phase difference.

As shown in Fig. 4-11, the tunnel diode is tested by applying a controlled d-c voltage across the diode, through resistor R_3. This d-c voltage is developed by rectifier CR_1, and is controlled by the variac. Current through the tunnel diode also flows through R_3. The voltage drop across R_3 is applied to the vertical channel, and causes the spot to move up and down. Therefore, vertical deflection is proportional to current. Vertical scale divisions can be converted directly to current when R_3 is made 100 ohms. For example, a 3-volt vertical deflection indicates 30 milliamperes.

The same voltage applied across the tunnel diode is applied to the horizontal channel (which has been voltage calibrated), and causes the spot to move from left to right. (For a tunnel diode test, the horizontal and vertical zero reference points should be at the *lower left* of the screen rather than in the center.) The horizontal deflection is proportional to voltage. The combination of the horizontal (voltage) deflection and vertical (current) deflection causes the spot to trace out the complete negative resistance characteristic.

1. Connect the equipment as shown in Fig. 4-11.

2. Place the oscilloscope in operation. Voltage calibrate both the vertical and horizontal channels as necessary. The spot should be at the lower left-hand side of center with no signal applied to either channel.

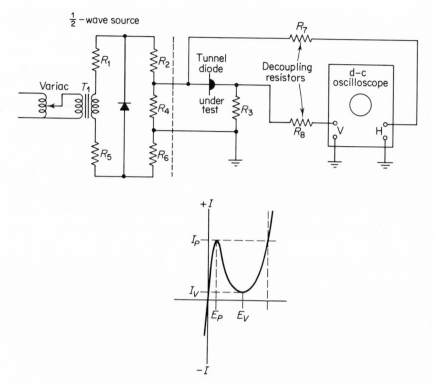

Fig. 4-11. Testing tunnel diode negative resistance characteristics.

3. Switch off the internal recurrent sweep. Set sweep-selector and sync-selector to external. Leave the horizontal- and vertical-gain controls set at the (voltage) calibration position.

4. Adjust the variac so that the voltage applied across the tunnel diode under test is the maximum-rated forward voltage. This can be read across the voltage-calibrated horizontal axis.

5. Check the oscilloscope pattern against the typical curve of Fig. 4-11, or against the tunnel diode manufacturer's data.

6. The following equation can be used to obtain a *rough approximation* of negative resistance in tunnel diodes:

$$\text{Negative resistance} = \frac{E_v - E_p}{2(I_p - I_v)} \qquad \text{where}$$

$E_v =$ Valley voltage

$E_p =$ Peak voltage

$I_v =$ Valley current

$I_p =$ Peak current

4-9. Diode Color Coding

In semiconductor component designations, the number-letter combination 1N indicates a diode. Standard color coding supplies the missing numbers and letters which identify the complete diode type.

Figure 4-12 shows the various diode color coding arrangements.

For the earlier type of diode which had two center digits (such as the 1N34A), three color bands are used as shown in Fig. 4-12(a). The color codes are grouped toward the cathode end and, for the diode shown, are read from left to right. The first two color code bands indicate the significant figures, and the third color band identifies the letter.

For diodes having three central digits (such as the 1N100A), four color bands are used as shown in Fig. 4-12(b). The first three bands from left to right identify the first three significant digits, and the last band provides the letter designation.

When the diode has four central digits (such as the 1N3666A), five color bands are used as shown in Fig. 4-12(c). The first four bands from left to right identify the first four significant digits, and the last band provides the letter designation.

Figure 4-12(d) shows an example of diode color coding where there are three digits followed by a letter (1N445B).

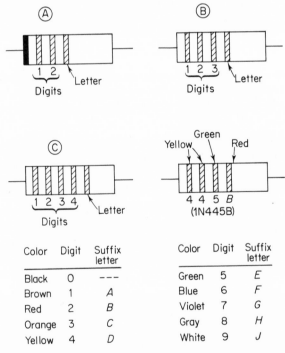

Color	Digit	Suffix letter
Black	0	---
Brown	1	A
Red	2	B
Orange	3	C
Yellow	4	D

Color	Digit	Suffix letter
Green	5	E
Blue	6	F
Violet	7	G
Gray	8	H
White	9	J

Fig. 4-12. Diode color codes.

CHAPTER 5

Semiconductor Power Supplies

This chapter is devoted to power supply circuits using semiconductor components. The major function of any power supply is to provide a source of d-c voltage for operation of other circuits. Usually, this involves rectifying alternating current into pulsating direct current, filtering the pulsating d-c into pure direct current, and then regulating the direct current to maintain a constant voltage and/or current. In the case of some equipment (usually portable), the process may also include converting direct current (from a battery) into alternating current so that it may be stepped up in voltage, then restored back to direct current at some higher voltage level.

The processes of rectification, regulation, and conversion can be accomplished using semiconductor components.

5-1. Basic Half-Wave Power Supply

The basic process of rectification by a diode is discussed in Chapter 1. When a single diode is used to convert alternating current into direct current, the circuit is known as half-wave since only one-half of each cycle is used. Usually a filter circuit is required with a half-wave rectifier to smooth out the pulsating d-c into pure d-c or, (more realistically) into d-c with a slight "ripple."

A basic half-wave diode rectifier and filter circuit are shown in Fig. 5-1. The diode CR_1 cathode element is connected to a filter capacitor C_1 and to a filter choke L_1. Another filter capacitor C_2 is also used to smooth out the

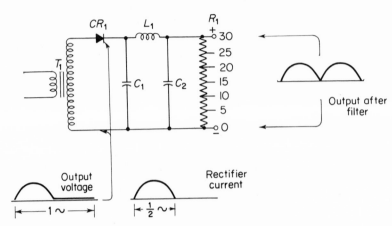

Fig. 5-1. Basic single-phase, half-wave rectifier circuit.

pulsating direct current. (A more detailed discussion of filter circuits is contained in the author's *Databook for Electronic Technicians and Engineers,* Prentice-Hall, Inc., 1968.) A resistance R_1 is placed across the power supply output. This resistance is not used in all circuits. Instead, the output is connected directly to the load circuit. If resistor R_1 is used, it is known as a *bleeder* resistor, since it places a small current drain on the power supply, and helps to stabilize the output.

If resistance R_1 is made up of a single tapped resistor, or a series of several resistors, it is possible to take several different voltages from the bleeder network. For example, if the power supply output was 30 volts, and the bleeder was made up of six equal-value resistors, the available voltages would be 5, 10, 15, 20, 25, and 30 volts respectively, as shown in Fig. 5-1.

When a positive alternation occurs on the diode CR_1 anode, current flows from one side of T_1 to bleeder R_1. Transformer T_1 is used to step up (or step down) the a-c line input voltage to some value, suitable for the circuits with which the power supply is to operate. Usually, T_1 is used to step the voltage *down* since most semiconductor circuits operate at voltages lower than the common power of 115 volts.

Initially, the filter capacitors do not have a charge. However, during the positive alternation, the capacitors charge to some value near the peak voltage of the transformer secondary. As the capacitors are charged, current flows through R_1 and choke L_1 to the cathode side of diode CR_1. The current is returned to the transformer through CR_1.

When the negative alternation occurs, the diode side of the secondary winding will be negative and the ground side will be positive. Because diode CR_1 permits current flow in one direction only, there will be no current flow from the transformer during the negative half-cycle. However, the filter capacitors will discharge, with current flowing from the negative or ground side to the positive side, through the bleeder resistor R_1. Therefore, the bleeder resistor has current flowing through it on both the positive and negative alternations or half-cycles, and a voltage is present across the bleeder at all times.

In Fig. 5-1, the bleeder-resistor voltage (or power-supply output voltage) is a direct current, but not necessarily a constant voltage. In practical applications, there will be some "ripple" or variation in amplitude. The output voltage decreases slightly between cycles (negative peak of the ripple voltage), and then increases at the peak of each half-cycle (positive peak of the ripple voltage). Usually, the peak-to-peak ripple voltage is expressed as a percentage of the total power-supply output voltage. For example, if the power supply produces 100 volts of direct current across the bleeder resistor, and the ripple is 3 volts peak-to-peak, there is a 3 per cent ripple. The amount of ripple for any given type of power supply circuit (half-wave, full-wave, etc.) depends upon the type of filter.

5-2. Full-Wave Power Supply

Both the positive and negative alternations of the a-c cycle are used in a full-wave supply. For this reason, the full-wave supply is more efficient than the half-wave circuit. However, the full-wave circuit requires two diodes and a transformer with a center-tap.

As shown in Fig. 5-2, one lead of the transformer secondary is connected to diode CR_1, while the opposite end of the secondary is connected to diode CR_2. The center-tap is connected to the common or ground circuit. The total voltage across the secondary of the transformer is about *twice* the voltage that appears at the power supply output.

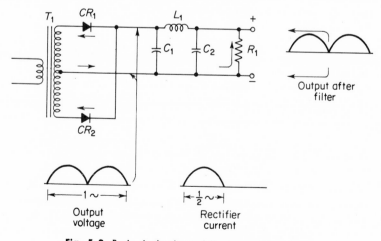

Fig. 5-2. Basic single-phase, full-wave rectifier circuit.

When a positive alternation occurs on the diode CR_1 anode, current flows from the center-tap of T_1, through the bleeder resistor R_1 and choke L_1, to the transformer secondary-winding end.

On the next alternation, when the anode of diode CR_2 is positive, current flows from the center-tap of T_1, through the bleeder resistor R_1 and choke L_1, to the opposite transformer secondary-winding end.

The filter capacitors charge at a rate which is *twice* that for half-wave rectification, since current flows through the bleeder resistor in the same direction on both alternations or half-cycles. Consequently, the ripple frequency is *twice* that of half-wave rectification ripple frequency.

Since the discharge time between the peaks of pulsating d-c is only half that found in half-wave rectification, because there is a shorter time between peaks, the filter capacitors do not have as long to discharge. This makes it easier to maintain a relatively high charge, and makes the output "smoother" than that of a half-wave supply.

5-3. Full-Wave Bridge Power Supply

A bridge circuit makes it possible to have full-wave rectification using a transformer without a center-tap. As shown in Fig. 5-3, four rectifier diodes are required in the basic bridge circuit.

When a positive alternation occurs, with the top end of the transformer secondary positive, current flows from the secondary bottom, through diode CR_3, the load or bleeder resistor R_1 and diode CR_2 to the secondary top.

On the next alternation, when the bottom end of the transformer secon-

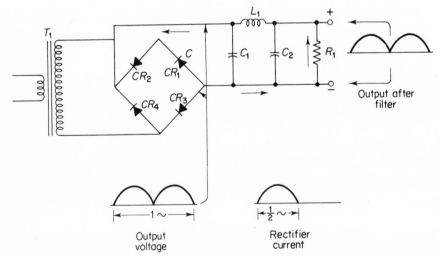

Fig. 5-3. Basic single-phase, full-wave, bridge rectifier circuit.

dary is positive, current flows from the secondary top, through diode CR_1, the load or bleeder resistor R_1, and diode CR_4 to the secondary bottom.

Full-wave rectification occurs since current flows through the bleeder or load resistance in the same direction on both half-cycles. This produces both the higher ripple frequency and more efficient filtering.

5-4. Three-Phase Power Supplies

The three-phase power supplies shown in Figs. 5-4 through 5-7 are similar to the basic half-wave and full-wave circuits, except that each phase is provided with its own diode (for half-wave) or diodes (for full-wave). This reduces the amount of current through each diode (making it possible to use smaller diodes), and increases the ripple frequency (making it easier to filter).

The three-phase Y half-wave circuit of Fig. 5-4 uses three rectifiers. One-third of the total output current flows through each rectifier, and the ripple frequency is three times that of a corresponding half-wave. That is, the ripple frequency is three times the line frequency.

The three-phase full-wave circuit of Fig. 5-5 uses six rectifiers. This circuit delivers twice as much voltage output as the circuit of Fig. 5-4 for the same transformer conditions. One-third of the total output current flows through

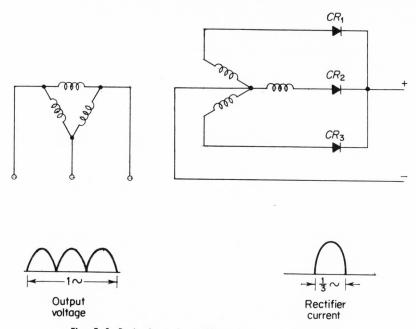

Fig. 5-4. Basic three-phase "Y" half-wave rectifier circuit.

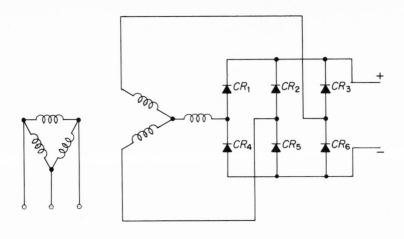

Output
voltage

Rectifier
current

Fig. 5-5. Basic three-phase "Y" full-wave rectifier circuit.

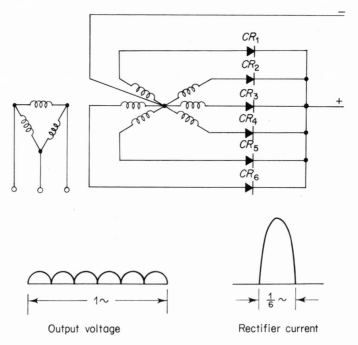

Output voltage

Rectifier current

Fig. 5-6. Basic six-phase "star" rectifier circuit.

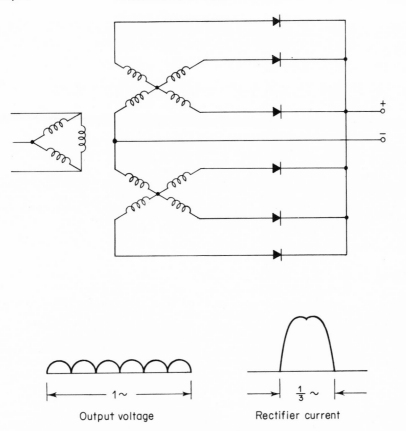

Output voltage   Rectifier current

Fig. 5-7. Basic three-phase "double-Y" and inter-phase transformer
rectifier circuit.

each rectifier, and the ripple frequency is three times that of a corresponding full-wave (or six times the line frequency).

The six-phase "star" circuit of Fig. 5-6 also uses six rectifiers. However, only one-sixth of the total output current flows through each rectifier.

The three-phase, double-Y, and interphase transformer circuit of Fig. 5-7 uses six half-wave rectifiers connected in parallel. This provides twice the output current of the half-wave circuit shown in Fig. 5-4, as well as a ripple frequency six times the line frequency.

5-5. Voltage/Current Ratios of Power Supply Circuits

Table 5-1 summarizes the voltage and current ratios of the power supply circuits shown in Figs. 5-1 through 5-7. When a particular rectifier type has been selected for use in a specific circuit, Table 5-1 can be used to determine the parameters and characteristics of the circuit.

TABLE 5-1

Voltage/Current Ratios for Semiconductor Power Supply Circuits

Circuit Ratios	Fig. 5-1	Fig. 5-2	Fig. 5-3	Fig. 5-4	Fig. 5-5	Fig. 5-6	Fig. 5-7
Output voltage:							
Average	E_{av}	E_{av}	E_{av}	E_{av}	E_{av}	E_{av}	E_{av}
Peak ($\times E_{av}$)	3.14	1.57	1.57	1.21	1.05	1.05	1.05
RMS ($\times E_{av}$)	1.57	1.11	1.11	1.02	1.00	1.00	1.00
Ripple (%)	121	48	48	18.3	4.3	4.3	4.3
Input voltage (RMS):							
Phase ($\times E_{av}$)	2.22	1.11*	1.11	0.855†	0.428†	0.740†	0.855†
Line-to-line ($\times E_{av}$)	2.22	2.22	1.11	1.48	0.74	1.48#	1.71**
Average output (load) current	I_{av}	I_{av}	I_{av}	I_{av}	I_{av}	I_{av}	I_{av}
Rectifier Cell Ratios							
Forward current:							
Average ($\times I_{av}$):	1.00	0.5	0.5	0.333	0.333	0.167	0.167
RMS ($\times I_{av}$): resistive load	1.57	0.785	0.785	0.587	0.579	0.409	0.293
Inductive load	—	0.707	0.707	0.578	0.578	0.408	0.289
Peak ($\times I_{av}$): resistive load	3.14	1.57	1.57	1.21	1.05	1.05	0.525

TABLE 5-1   (cont'd)

Inductive load	—	1.00	1.00	1.00	1.00	1.00	0.500
Ratio peak-to-average: resistive load	3.14	1.57	1.57	1.21	1.05	1.05	0.525
Inductive load	—	2.00	2.00	3.00	3.00	6.00	3.00
Peak reverse voltage: $\times E_{av}$	3.14	3.14	1.57	2.09	1.05	2.42	2.09
$\times E_{rms}$	1.41	2.82	1.41	2.45	2.45	2.83	2.45

* to center tap
† to neutral
≠ maximum value
**maximum value, no load

The following conditions must be applied to the data contained in Table 5-1:

The values apply for *resistive* and *inductive* loads, not to capacitive loads. Usually, the half-wave circuit of Fig. 5-1 is used with a resistive load. The remaining circuits of Figs. 5-2 through 5-7 are used with inductive loads.

Current ratios given for inductive loads apply *only when a filter choke* is used between the output of the rectifier and any capacitor in the filter circuit.

The values *do not* take into account voltage drops in the power transformer, rectifiers, or filter components under load conditions.

All ratios are shown as functions of either the average output voltage E_{av}, or the average d-c output current I_{av}, both of which are expressed as unity for each circuit.

5-6. Voltage Doubling and Tripling Circuits

Voltage doubling and tripling are used in solid-state power-supply circuits where a higher voltage is required. Using such circuits, it is possible to increase an available a-c voltage *without* a transformer.

Voltage Doubling Circuit

As shown in Fig. 5-8, the basic voltage doubling circuit requires two rectifier diodes and two capacitors. When the a-c line alternation is such that

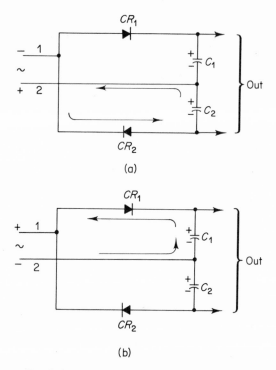

Fig. 5-8. Basic voltage doubling rectifier circuit.

terminal 1 is negative, and terminal 2 is positive, as shown in (a) of Fig. 5-8, current will flow from terminal 1 in the direction shown by the arrows. The path of current from terminal 1 includes diode CR_2, and capacitor C_2, then returns to the positive terminal 2. Capacitor C_2 is charged during this current flow with a polarity as shown.

During the next alternation, when terminal 1 is positive and terminal 2 is negative, current will flow as shown in (b) of Fig. 5-8. As a result, current flows from terminal 2 toward C_1, and charges capacitor C_1 to the *peak* of the a-c voltage. Current continues through diode CR_1 to the positive terminal 1.

Capacitors C_1 and C_2 are charged to the peak values of the a-c line voltage on alternate half-cycles. The d-c output voltage is taken from across the *two capacitors*. Therefore, the d-c voltage will be approximately double that of the a-c voltage. The two capacitors are, in effect, in series, with the polarities adding to increase the total voltage.

Voltage Tripling Circuit

As shown in Fig. 5-9, the basic voltage-tripling circuit requires three rectifier diodes and three capacitors. When the a-c line alternation is such

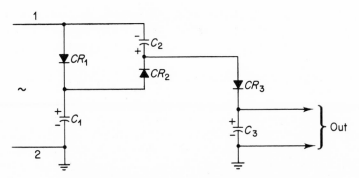

Fig. 5-9. Basic voltage tripling rectifier circuit.

that terminal 1 is positive, and terminal 2 is negative, current flows from terminal 2 through CR_1 to terminal 1. Diode CR_1 will be forward biased under this condition, permitting capacitor C_1 to be charged to the a-c line voltage.

During the next alternation, when terminal 1 is negative and terminal 2 is positive, diode CR_1 will be reverse biased, and will not conduct. Instead, diode CR_2 will be forward biased, and permit current flow toward terminal 2, via capacitor C_1. The charge which is already across C_1 will add to the a-c line voltage, and place a charge across C_2. This charge will be proportional to the line voltage added to the charge already across C_1.

During the next alternation, when terminal 1 is again positive, and terminal 2 is negative, capacitor C_1 is again charged as before. But, at the same time, diode CR_3 conducts, and charges capacitor C_3. When diode CR_3 conducts, the charge across capacitor C_3 will be composed of the line voltage, plus the existing charge across capacitor C_2. Since the capacitor C_2 charge is already *double* voltage, this is added to the line voltage, and produces a *triple* voltage across capacitor C_3.

5-7. Regulation Circuits

The amount of voltage that can be obtained from a power supply is determined by the voltage available at the source (line or transformer), less any voltage drop through the rectifiers and filter. This voltage drop is determined by the amount of current flowing through the power supply components (rectifiers, chokes, etc.).

As more current is drawn from the power supply, the voltage drop across the power-supply components increases, and the output voltage is reduced. When less current is drawn from the power supply, the output will be increased, since the power-supply components will have a minimum voltage drop.

The variation of voltage output with respect to the amount of current drawn from the power supply is known as *voltage regulation*, or simply *regulation*. Usually, regulation is expressed as a percentage. The value is determined by:

$$\text{percentage regulation} = \frac{(\text{no-load voltage}) - (\text{full-load voltage})}{(\text{full-load voltage})} \times 100$$

This equation applies to voltage regulation, and takes into account the proportions of voltage increase and decrease with a change of load on a power supply. The smaller the difference between the full-load and no-load voltages, the better the regulation.

The filter components (chokes and capacitors) provide some measure of regulation for the power-supply circuit. Should the output voltage drop, the capacitors will discharge, and serve to keep the voltage constant. Likewise, a voltage drop will cause the magnetic field around the filter chokes to collapse (or decrease). The moving magnetic lines of force produce a current in the coil, but in the opposite direction to the current flow causing the lines of force. Thus, the filter choke sets up an opposition to any change in voltage amplitude. The regulation provided by the filter network is sufficient for many appliciations. In other circumstances, the power supply output must be maintained at some critical voltage or current. A number of semiconductor circuits have been developed to provide both voltage and current regulation. The following paragraphs of this section summarize these circuits:

Zener Diode Regulation

The most common method of providing voltage regulation for semiconductors is with a Zener diode. As discussed in previous chapters, a Zener diode will maintain the voltage drop across its terminals at constant value, no matter what current is being drawn.

In its simplest form the Zener diode regulator consists of a series resistance R_s and a shunt-connected diode CR_1 as shown in Fig. 5-10. The value of R_s is set by the load requirements. If R_s is too large, the diode will not be able to regulate at large values of load current (I_L). If R_s is too small, the diode dissipation rating (power or wattage rating) may be exceeded at low I_L values.

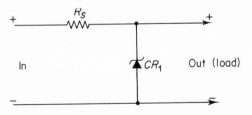

Fig. 5-10. Basic Zener diode regulator circuit.

The optimum value for R_s can be calculated by:

$$R_s = \frac{\text{Minimum supply voltage} - \text{Zener voltage}}{\text{Maximum load current} + 0.1 \text{ of maximum load current}}$$

If the value of R_s is known, the maximum diode dissipation can be calculated by:

$$\begin{matrix} \text{Zener} \\ \text{diode} \\ \text{dissipation} \end{matrix} = \left[\frac{\begin{matrix} \text{Maximum} \\ \text{supply} \\ \text{voltage} \end{matrix} - \begin{matrix} \text{Zener} \\ \text{voltage} \end{matrix}}{R_s} - \begin{matrix} \text{Load} \\ \text{current} \end{matrix} \right] \times \begin{matrix} \text{Zener} \\ \text{voltage} \end{matrix}$$

Sometimes it is necessary to regulate a voltage not available in standard Zener diodes. This condition can be overcome by various circuit arrangements.

Several Zener diodes can be connected in series as shown in Fig. 5-11. The total regulated voltage will be the sum of the individual Zener voltages. For example, if the Zener voltages of the diodes shown in Fig. 5-11 were 5, 10, and 15 volts respectively, the total regulated output would be 30 volts. The diodes need not have equal breakdown voltages since the arrangement is self equalizing. However, the *power handling* ability of each diode should be the same. Likewise, the *current ranges* should be similar or the loads so arranged to avoid damaging any of the diodes.

The diode can also be used as a series element as shown in Fig. 5-12. This circuit is to be used where only a small voltage drop is required. The

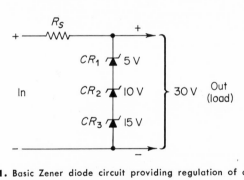

Fig. 5-11. Basic Zener diode circuit providing regulation of a voltage higher than individual Zener voltages.

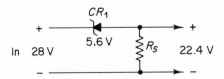

Fig. 5-12. Basic Zener diode series circuit to provide regulation where small voltage drops are required.

shunt circuit of Fig. 5-11 is used for large voltage drops. In the series circuit of Fig. 5-12, the drop across the diode is 5.6 volts, lowering the 28-volt input to 22.4 volts. It should be noted that the entire load current, plus the current through R_s, passes through the series diode. Thus, the total current must be used in calculating the required power handling ability (wattage rating) of the diode.

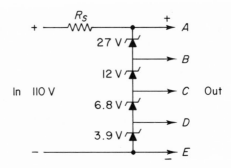

Fig. 5-13. Basic Zener diode circuit used as a voltage divider.

Several Zener diodes can be connected in series to act as a *divider* to obtain several regulated voltages as shown in Fig. 5-13. Using the four diodes shown, it is possible to obtain the following 10 voltages:

VOLTAGE	TERMINALS
3.9	D—E
6.8	C—D
10.7	C—E
12.0	B—C
18.8	B—D
22.7	B—E
27.0	A—B
39.0	A—C
45.8	A—D
49.7	A—E

It is possible to arrange Zener diodes so as to produce a regulated output at a voltage *lower* than that of the diode. Such a circuit is shown in Fig. 5-14 where two diodes are used. The output is the regulated *difference* voltage $(8.2 - 6.8 = 1.4)$. This arrangement provides good temperature compensation since both diodes tend to drift in the same direction, maintaining the difference voltage.

Zener diodes can also be arranged to provide an *adjustable* regulated voltage as shown in Fig. 5-15. Any combination of diodes can be used to obtain the desired range of output voltage.

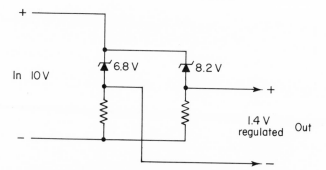

Fig. 5-14. Basic Zener diode circuit ueed as a difference supply for low voltages.

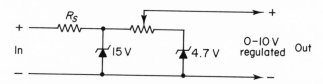

Fig. 5-15. Basic Zener diode circuit used to provide an adjustable regulated output.

Regulation Effect

A perfect Zener diode (or any regulator circuit) should produce a *constant output* with any change in *input voltage* or with *load current*. This is not possible under practical conditions.

The ability of a regulator to absorb input voltage variations is expressed by:

$$\text{Input regulation factor} = \frac{\Delta V_o V_{in}}{\Delta V_{in} V_o} = \text{with load current constant}$$

where Δ = change in voltage

The ability of a regulator to absorb output current variations is expressed by:

$$\text{Load regulation factor} = \frac{\Delta V_o R_L}{\Delta R_L V_o}$$

where ΔV_o = change in voltage

ΔR_L = change in load resistance

Both the input and load regulation factors can be considered as a figure of merit for the complete power supply, since they express the regulating ability of the supply, not just the regulating circuit. In both cases, the figures of merit should be the *minimum*.

Output Impedance

The output impedance of a regulated power supply (using a Zener diode, or complete regulator circuit) should be low. In theory, the output impedance should be zero.

The d-c output resistance is the ratio of change in output voltage to change in output current with the input voltage held constant, as expressed by:

$$\text{D-c output resistance} = \frac{\Delta V_o}{\Delta I_L}$$

where ΔV_o = change in voltage
ΔI_L = change in load current

The dynamic output impedance is the ratio of the a-c components of the output voltage and current when the load is varied sinusoidally with the input voltage held constant.

Power-supply output impedance increases with decreases in load current.

Extending Zener Regulation

The voltage control ability of a Zener diode can be increased if the diode is used to control the operating point of a transistor, or group of transistors. There are two basic types of transistor regulators: *shunt* and *series*. The shunt regulator is placed across the power supply output, while the series regulator is placed in series with the power supply output.

Shunt Voltage Regulator

The simplest form of shunt-transistor voltage regulator is shown in Fig. 5-16. Transistor Q_1 appears across the power supply output as a variable "bleeder" resistor, with current flowing between emitter and collector. Base current flows through Zener diode CR_1. Both of these currents, as well as the load current, flow through the series resistor R_1.

When the forward bias of the Q_1 base-emitter circuit is increased, the emitter-collector current is also increased (the emitter-collector resistance

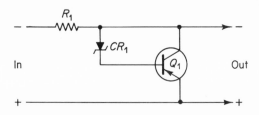

Fig. 5-16. Basic shunt regulator circuit.

across the output is lowered). Less current is drawn through resistor R_1, producing a greater voltage drop across R_1. In turn, this lowers the output voltage.

If the power supply load should increase, more current will be drawn through the series resistor R_1, and the power-supply output voltage will drop. Under these conditions, less current will be drawn through CR_1. Since less current will then be drawn through the base-emitter of Q_1, the forward bias is lowered, and less emitter-collector current is drawn through from the power supply. This causes a lower drop across series resistor R_1, which tends to increase the power-supply output voltage, and thus offset the initial drop in voltage.

When the power supply is subject to large current changes, shunt regulators are often connected in *cascade* to increase their effectiveness. A typical cascade shunt regulator is shown in Fig. 5-17. Transistors Q_1 and Q_2 are placed across the power supply output, and act as variable resistors. Base current for Q_1 flows through Zener diode CR_1. Base current for Q_2 flows through emitter resistor R_2. The voltage across R_2 is determined by current flowing through R_2. All of these currents, as well as the load current, flow through the series resistor R_1.

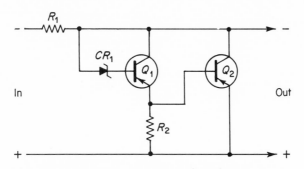

Fig. 5-17. Basic cascade shunt regulator circuit.

As power supply load decreases, less current is drawn through R_1, and the output voltage rises. More current will be drawn through CR_1, and through the base-emitter of Q_1. This raises the forward bias of Q_1, causing more emitter-collector current to be drawn from the power supply. Also, more current is passed through emitter resistor R_2, causing the voltage drop across R_2 to increase. The forward bias on Q_2 increases, and more Q_2 emitter-collector current is drawn from the power supply. The increased current through Q_1 and Q_2 causes a higher drop across R_1, which tends to decrease the power-supply output voltage, and thus offset the initial rise in voltage.

The transistor shunt regulator can be adapted to supply voltage *lower* or *higher* than the supply voltage.

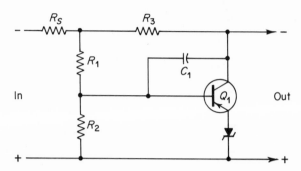

Fig. 5-18. Basic shunt regulator for voltages higher than the Zener voltage.

Figure 5-18 shows a shunt regulated supply for outputs *higher* than the Zener voltage. Neglecting the effect of R_s, or assuming that the source voltage is at the junction of R_s and R_1, the output voltage will be determined by the ratio of R_1/R_2, or:

$$\frac{\text{Output}}{\text{voltage}} = \frac{R_1 + R_2}{R_2}$$

For example, if R_1 and R_2 are of the same value, the output voltage will be twice that of the Zener voltage.

Resistor R_3 compensates for variations in the supply to the regulator circuit. A high resistance of R_3 will provide overcompensation, while an R_3 with too low a resistance will produce undercompensation. The value of R_3 is often found by trial and error, using a variable resistance.

Figure 5-19 shows a shunt regulated supply for outputs *lower* than the Zener voltage. Potentiometer R_2 acts as a variable voltage divider, and sets the regulated output voltage. The forward bias on Q_1 is determined by the voltage drop across R_1 (or the current through R_1). If the supply voltage tends to increase, more current flows through CR_1 and R_1, causing an increase in forward bias on Q_1. The increase in Q_1 collector-emitter current passing through R_s causes a greater voltage drop across R_s, offsetting the initial rise in supply voltage.

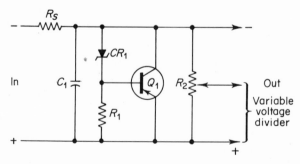

Fig. 5-19. Basic shunt regulator for voltages lower than the Zener voltage.

Series Voltage Regulator

The series circuit is usually used where higher currents must be regulated. The basic series transistor voltage regulator is shown in Fig. 5-20. Transistor Q_1 is placed in series with the power supply output, together with series resistor R_1. Transistor Q_1 acts as a variable "series resistor" with current flowing from the collector to emitter. Current also flows through resistor R_2 and the Zener diode CR_1 to establish a voltage at the base of Q_1. The Q_1 base voltage is determined by current flow through R_2. The base voltage remains fixed in relation to the positive terminal of the power supply, but will vary in relation to the negative terminal.

When the forward bias of the base-emitter circuit is increased, the emitter-collector "resistance" in series with the power-supply output is lowered. This causes a lower voltage drop across the emitter-collector resistance, and raises the power-supply output voltage.

For example, if the power-supply load should increase, more current will be drawn through the series resistor, as well as the Q_1 emitter-collector resistance, and the power-supply output voltage will drop. Under these conditions, less current will be drawn through CR_1 and R_2, increasing the forward bias on Q_1. In turn, the decrease in Q_1 emitter-collector resistance raises the power-supply output voltage.

Although semiconductors are not usually associated with high voltages, it is possible to use semiconductor circuits to regulate high-voltage power supplies. Such a circuit is shown in Fig. 5-21. Transistor Q_1 is placed in series with the power-supply output, and acts as a variable series resistor. Current also flows through R_1 to establish a voltage at the base of Q_1. This voltage is determined by the current flowing through resistor R_1, and through the emitter-collector circuit of Q_2.

When the forward bias of Q_1 is decreased, the Q_1 emitter-collector resistance in series with the power supply is raised. This causes a higher voltage drop across the emitter-collector resistance, and lowers the power-supply output voltage.

The emitter-collector current of Q_2 is determined by the forward bias voltage obtained across R_2. This voltage is determined by the emitter-collector current of Q_3. The emitter of Q_2 is held at a fixed voltage by the action of Zener diode CR_1.

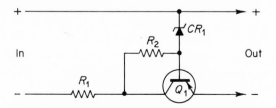

Fig. 5-20. Basic series regulator circuit.

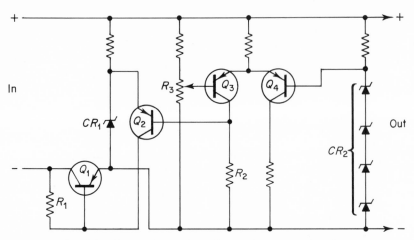

Fig. 5-21. Semiconductor circuit for regulation of high voltages.

Transistors Q_3 and Q_4 are used to establish a control voltage for the regulation circuit. The base of Q_4 is held at a fixed voltage by the action of Zener diode CR_2. The emitter-collector current of Q_4 remains constant, and the emitter of Q_4 remains at a fixed voltage. Since the emitter of Q_3 is connected directly to the emitter of Q_4, the emitter of Q_3 also remains at a fixed voltage. However, the voltage at the base of Q_3 varies with the power-supply output voltage.

If the power-supply voltage should change, the forward bias on Q_3 will be changed, causing a change in the voltage drop across R_2. This changes the foward bias on Q_2, causing a change in emitter-collector current. Since the emitter-collector current of Q_2 passes through R_1, the voltage drop across R_1 changes, making the base of Q_1 more or less negative. This changes the forward bias on Q_1 and changes the emitter-collector resistance in series with the power-supply output. In turn, this causes a higher or lower voltage drop across the emitter-collector resistance, and changes the power-supply output voltage to offset the initial rise or fall in voltage.

The power-supply output voltage is set by adjustment of potentiometer R_3 which determines the bias voltage on the base of Q_3. In practice, the power supply is adjusted by connecting a voltmeter across the power-supply output terminals (usually with the load connected), and setting R_3 for the desired voltage.

High Current Regulators

If a very high current must be regulated, and all of this current passed through a single series transistor, the transistor might not be able to dissipate all of the heat satisfactorily, even though heat sinks were used. This could result in damage to the transistor. To overcome the problem, several transistors can be connected in parallel with each other. Then the string of

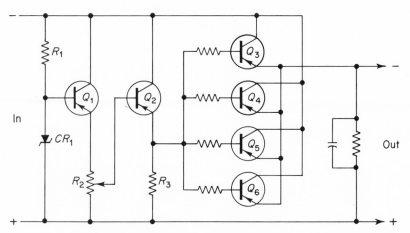

Fig. 5-22. Semiconductor circuit for regulation of voltages where high currents are present.

transistors is connected in series with the power-supply output, as shown in Fig. 5-22.

Transistors Q_3 through Q_6 are connected in series with the power-supply output, and in parallel with each other, to act as variable series "resistors." Current is divided equally between all four so that each transistor dissipates one-fourth of the total heat.

The base of Q_1 is held at a fixed voltage (in relation to the emitter) by the action of Zener diode CR_1. However, the voltage across the emitter-collector circuit of Q_1 varies with the power-supply voltage. If the power-supply voltage changes for any reason, the voltage across R_2 varies, and changes the forward bias on Q_2. In turn, this changes the current through R_3. Since the voltage across R_3 determines the forward bias on all four transistors (Q_3 to Q_6), the voltage across R_3 determines the amount of resistance offered by transistors Q_3 to Q_6. A change in Q_3 to Q_6 voltage drop changes the power-supply output voltage to offset the initial change in voltage. The power-supply output voltage is set by adjustment of R_2 which determines the forward bias on the base of Q_2.

Constant Current Regulation

Semiconductors can be used to regulate a power supply so as to produce a *constant current*, rather than a constant voltage. The basic circuit is shown in Fig. 5-23.

Transistor Q_1 functions as a variable "series resistor" in the power-supply output. Two parallel circuit paths exist. One path is through Zener diode CR_1 in series with bias resistor R_3. The other path is through resistor R_1 and transistor Q_1.

Should there be any variation in current at the power-supply output, the

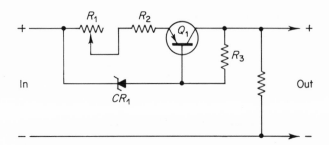

Fig. 5-23. Basic semiconductor current regulator circuit.

current through R_3 will vary, and change the forward bias on Q_1. In turn, the emitter-collector resistance of Q_1 will vary to correct the current flow. The net result is that for every change in R_3 current, there is an equal and opposite change in Q_1 current.

The current output of such a circuit is set by R_1. The current remains constant (within limits) in spite of any load changes. However, the power-supply output voltage will vary and the load changes.

5-8. Power Conversion Circuits

Semiconductors can be used effectively in power-conversion circuits. Semiconductor circuits have replaced the vibrator-type circuits for converters and inverters. Usually, the term *inverter* refers to a device used to change d-c power into a-c power. If the a-c output is rectified and filtered to provide d-c (say at a higher voltage), the over-all circuit is referred to as a *converter*.

The *push-pull switching inverter* is the most widely used type of power-conversion circuit. For inverter applications, the circuit provides a square-

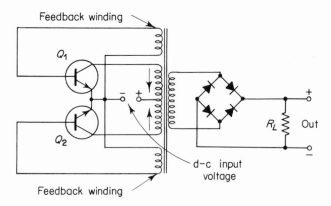

Fig. 5-24. Basic push-pull semiconductor converter circuit with single transformer.

wave a-c output. When the inverter is used to provide d-c to d-c conversion, the squarewave voltage is usually applied to a bridge rectifier and filter.

Figure 5-24 shows the basic push-pull converter circuit using a *single* transformer. When the d-c input voltage is applied to the circuit, current tends to flow through both transistors Q_1 and Q_2. Due to the normal unbalance of the circuit, one transistor will conduct more heavily than the other. Assuming that transistor Q_1 conducts more heavily (initially), the rise in current through its collector inductance (transformer winding) causes a voltage to be inducted in the feedback windings of transformer T_1, which supply the base drive to Q_1 and Q_2. The base voltages are in the proper polarity to increase the current through Q_1, and to decrease the current through Q_2. Transistor Q_1 is then quickly driven to saturation, which Q_2 is driven to cut off.

When Q_1 reaches saturation, the magnetic field around the collector is constant, and no feedback voltage is applied to the transistors. With the cutoff base voltage removed, current is allowed to flow through transistor Q_2. The increase in current through the collector inductance of Q_2 causes voltages to be induced in the feedback windings, in the polarity that increases the current through Q_2 and decreases the current through Q_1. This effect is aided by the collapsing magnetic field about the collector inductance of Q_1 that results from the decrease in current through Q_1. The feedback voltages produced by the collapsing field quickly drive Q_1 beyond cutoff, and further increase the current flow through Q_2 until saturation is reached. A new cycle is initiated each time one of the two transistors reaches saturation.

The squarewave voltage produced by the switching action of transistors Q_1 and Q_2 is coupled by transformer T_1 to the bridge rectifier and filter which develop a smooth, constant amplitude d-c voltage across the load resistance R_L. Push-pull transformer-coupled converters with full-wave rectification provide power to the load continuously and are well suited for low-impedance, high-power applications.

Figure 5-25 shows the basic push-pull converter circuit using *two* trans-

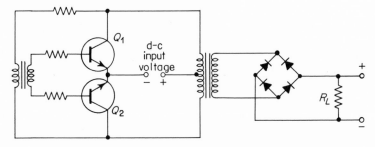

Fig. 5-25. Basic push-pull semiconductor converter circuit with two transformers.

formers. In this circuit, a small saturable transformer provides the base drive for the switching transistors, and a nonsaturable output transformer provides the coupling and desired voltage transformation (usually a step-up) of the output delivered to the load circuit.

With the exception that it uses a separate saturable transformer rather than feedback windings on the output transformer, to provide base drive for the transistors, the converter of Fig. 5-25 is very similar in operation to the circuit of Fig. 5-24.

CHAPTER 6

Unijunction Transistors

The unijunction transistor (or UJT) operates on an entirely different principle from that of the common (two-junction) transistor. The UJT is a *negative resistance* unit. That is, under the proper conditions, the input voltage or signal can be decreased, yet the output or load current will increase. Once the UJT is "turned on," it will not "turn off" until the circuit is broken, or the input voltage is removed. For this reason, the UJT makes an excellent trigger source. The UJT can be biased just below the "firing" point. When a small trigger voltage (either intermittant or constant) is applied, the UJT will "fire" and produce a large output-voltage pulse or signal that will remain on until the circuit is broken (by switching off the base voltage).

6-1. Operating Theory of Unijunction Transistors

The UJT is a three-terminal device. The three terminals are: emitter (E), base one (B_1) and base two (B_2). Table 6-1 shows the commonly used UJT symbols and their proper definitions in accordance with the JEDEC (Joint Electron Device Engineering Council) standard.

Figure 6-1 shows a simplified bar structure and equivalent circuit for a UJT. This bar structure was typical of the structures used in early UJTs, and is still in use today. The bar structure is used here (to explain operation of the UJT) for the sake of convenience. However, many different structures are used in the manufacture of UJTs, as is discussed in a later section.

TABLE 6-1

Unijunction Transistor Nomenclature

Symbol	Definition
I_E	Emitter current.
I_{EO}	Emitter-reverse current. Measured between emitter and base two at a specified voltage, and base one open circuited.
I_p	Peak-point emitter current. The maximum emitter current that can flow without allowing the UJT to go into the negative resistance region. Peak point is the lowest current on the emitter characteristic where: $$\frac{\Delta V_{EB_1}}{\Delta I_E} = 0$$
I_V	Valley-point emitter current. The current flowing in the emitter when the device is biased to the valley point. Valley point is the second lowest current on the emitter characteristic where: $$\frac{\Delta V_{EB_1}}{\Delta I_E} = 0$$
r_{BB}	Interbase resistance. Resistance between base two and base one, measured at a specified interbase voltage.
$V_{B_2 B_1}$	Voltage between base two and base one. Positive at base two.
V_p	Peak-point emitter voltage. The maximum voltage seen at the emitter before the UJT goes into the negative-resistance region.
V_D	Forward voltage drop of the emitter junction.
V_{EB_1}	Emitter to base one voltage.
$V_{EB_1(SAT)}$	Emitter saturation voltage. Forward voltage drop from emitter to base one at a specified emitter current (larger than I_V) and specified interbase voltage.
V_v	Valley-point emitter voltage. The voltage at which the valley point occurs with a specified $V_{B_2 B_1}$.
V_{OB_1}	Base-one peak-pulse voltage. The peak voltage measure across a resistor in series with base one when the UJT is operated as a relaxation oscillator in a specified circuit.
η	Intrinsic standoff ratio. Defined by the relationship: $$\eta = \frac{V_p - V_D}{V_{B_2 B_1}}$$
ar_{BB}	Interbase resistance temperature coefficient. Variations of resistance between B_2 and B_1 over the specified temperature range, and measured at the specified interbase voltage and temperature, with emitter open circuited.
I_{B_2}(mod)	Interbase modulation current. B_2 current modulation due to firing. Measured at a specified interbase voltage, emitter voltage, and temperature.

When voltage $V_{B_2 B_1}$ is applied, a current will flow in the silicon bar from base two to base one. Since the bar is essentially a resistor (r_{BB}), the current that flows into base two is determined by:

$$I_{B_2} = \frac{V_{B_2 B_1}}{r_{BB}} \tag{6-1}$$

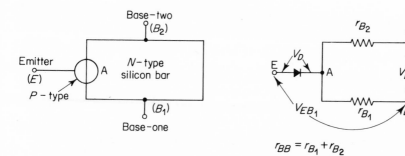

(a) A simplified bar structure   (b) Equivalent circuit valid for $I_E \leqq I_P$

Fig. 6-1. Simplified bar structure and UJT equivalent circuit (*courtesy Motorola*).

A fraction of the applied voltage $V_{B_2B_1}$ will appear at point A (Fig. 6-1), where the emitter is alloyed into the silicon bar. This fraction is the *intrinsic standoff* ratio η. The voltage at point A is therefore $\eta \times V_{B_2B_1}$. The *PN* junction formed by the emitter and the silicon bar is reverse biased, and only a small reverse leakage current flows in the emitter lead.

As the voltage V_E at the emitter is increased, a point will be reached where V_E equals the voltage at point A plus the forward voltage drop of the *PN* junction, V_D. The emitter voltage at this point is called the peak-point emitter voltage V_p. The peak-point voltage can be written:

$$V_p = V_D + \eta V_{B_2B_1} \tag{6-2}$$

The *PN* junction is now forward biased, and holes will be injected from the emitter into the silicon bar. The electric field inside the bar set up by $V_{B_2B_1}$ is of such a direction that the injected holes will be moved toward the base one terminal.

When the holes are injected into the bar from the emitter, an equal amount of electrons will be injected from base one to maintain charge neutrality. Since both the electron and hole concentrations increase in the silicon bar between the emitter and base one, the conductivity will also increase, and the resistance will decrease. This process is called *conductivity modulation*.

The decrease in resistance will cause a decrease in the voltage drop from emitter to base one, which in turn allows more holes to be injected from the emitter. This regenerative action continues until the UJT is in the so-called *negative resistance* region. An equivalent circuit for this region is shown in Fig. 6-2. At emitter currents equal to or less than I_p, the resistance r_{BB} can be divided into two parts: r_{B_1} and r_{B_2} according to the relations:

$$r_{B_1} = \eta r_{BB} \quad \text{and} \quad r_{B_2} = r_{BB} - r_{B_1} \tag{6-3}$$

In the negative resistance region, resistor r_{B_1} can be thought of as consisting of a fixed portion r_s, and a variable portion r_N, where r_s is the saturation

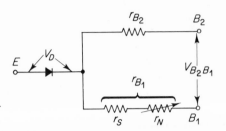

Fig. 6-2. Equivalent circuit for UJT in negative resistance region (*courtesy Motorola*).

resistance and r_N is the negative resistance, the magnitude of which decreases with increasing emitter current. When the negative resistance r_N decreases to zero, and r_{B_1} is equal to r_s only, the UJT is no longer in the negative resistance region.

The point on the emitter characteristic where r_{B_1} just reaches its minimum value is called the *valley point*. The emitter current and voltage at this point are the valley-point emitter-current I_V, and the valley-point emitter-voltage V_v, respectively.

When the emitter current is increased beyond I_V, the UJT enters the so-called *saturation region* where the emitter current is essentially a linear function of the emitter voltage. The equivalent circuit for the saturation region is shown in Fig. 6-3.

The standard unijunction transistor symbol with appropriate terms for current and voltage is given in Fig. 6-4. The base one and base two leads are shown at right angles to the base because they are nonrectifying (not a junction). However, the emitter connection is represented by an arrow because it is a rectifying *PN* junction, and the arrow is slanted to indicate the emitting properties of the junction. The arrowhead is pointing toward the base, showing a *P*-type emitter and an *N*-type base.

A static emitter-characteristic curve for a single value of $V_{B_2B_1}$ is shown

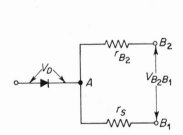

Fig. 6-3. Equivalent circuit for UJT in saturation region (*courtesy Motorola*).

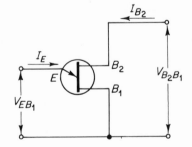

Fig. 6-4. UJT symbols and nomenclature (*courtesy Motorola*).

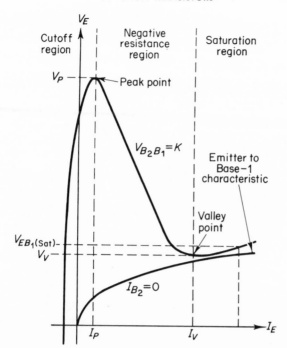

Fig. 6-5. Static UJT emitter characteristic curves (*courtesy Motorola*).

in Fig. 6-5. The emitter curve is not drawn to scale in order to show the different operating regions in more detail. The region to the left of the peak point is called the *cutoff region*. The emitter junction is reverse biased in most of the cutoff region, but is slightly forward biased at the peak point.

The region between the peak point and the valley point, where the emitter junction is forward biased and conductivity modulation takes place, is called the *negative resistance* region.

The region to the right of the valley point, where the emitter current is limited only by r_s, is called the *saturation region*.

The curve for base-two current (I_{B_2}) equal to zero is essentially the forward characteristic of a conventional silicon diode.

6-2. Structure of Unijunction Transistors

The bar structure (discussed in Sec. 6-1) was used frequently during the early development of the UJT. The *cube structure* was also used to some extent. Figure 6-6 is a cross-section diagram of the bar and cube structures.

The bar structure in Fig. 6-6(a) is formed by mounting a high resistivity N-type silicon bar on ceramic platform having an air gap in the center, and gold-antimony film deposited on each side of the gap. Base one and base

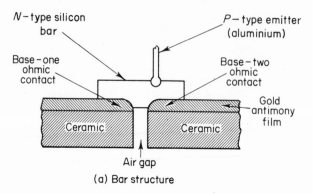

(a) Bar structure

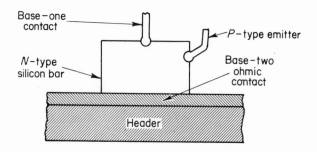

(b) Cube structure

Fig. 6-6. Cross-section diagram of UJT bar and cube structures (*courtesy Motorola*).

two are ohmic contacts that are formed between the silicon bar and the gold. A single *P*-type emitter is formed by alloying an aluminum wire onto the bar opposite from the base contacts.

The cube structure in Fig. 6-6(b) uses a high resistance *N*-type silicon cube. The cube is mounted on a header with a gold-antimony alloy contact between the bottom of the cube and the header. The base two ohmic contact is made to the gold antimony area. Base one is formed by alloying a gold wire to the top of the silicon bar. The emitter is formed in the same way by alloying an aluminum wire to the side of the cube.

Figure 6-7 shows the production steps of an *annular* UJT developed by Motorola. This process lends itself to modern automatic production methods, while retaining quality construction.

6-3. Characteristics of Unijunction Transistors

The *static emitter characteristics* shown in Fig. 6-5 were not drawn to scale, in order to show all three regions (cutoff, negative resistance, saturation)

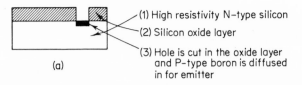

(1) High resistivity N-type silicon

(2) Silicon oxide layer

(3) Hole is cut in the oxide layer and P-type boron is diffused in for emitter

(a)

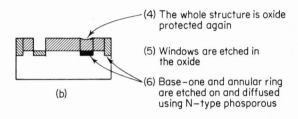

(4) The whole structure is oxide protected again

(5) Windows are etched in the oxide

(6) Base-one and annular ring are etched on and diffused using N-type phosporous

(b)

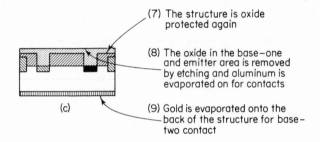

(7) The structure is oxide protected again

(8) The oxide in the base-one and emitter area is removed by etching and aluminum is evaporated on for contacts

(9) Gold is evaporated onto the back of the structure for base-two contact

(c)

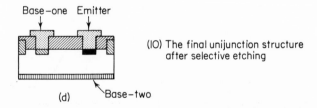

Base-one   Emitter

(10) The final unijunction structure after selective etching

(d)   Base-two

Fig. 6-7. Motorola annular UJT construction (*courtesy Motorola*).

of a UJT. Figures 6-8 and 6-9 show a typical UJT emitter curve for $V_{B_2B_1}$ = 20 volts, drawn to scale.

Figure 6-8 shows part of the cutoff region plotted on linear scale. When the emitter voltage is zero, the emitter current is negative. Peak-point voltage is reached at a forward emitter current of about 10 microamperes. As can be seen from Fig. 6-9, the voltage remains constant until the emitter current reaches approximately 0.1 microampere, at which point the voltage starts to decrease.

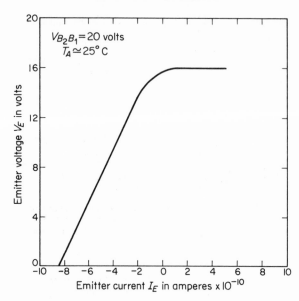

Fig. 6-8. UJT static emitter characteristics in cutoff region (*courtesy Motorola*).

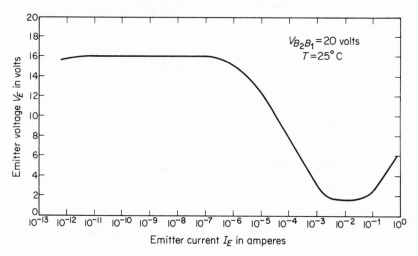

Fig. 6-9. UJT static emitter characteristics in negative resistance region (*courtesy Motorola*).

Peak current I_p for the UJT graphed in Figs. 6-8 and 6-9 therefore equals 0.1 microampere, while peak voltage V_p, equals 16 volts. Referring to Fig. 6-9, valley voltage, V_v, can be seen to be about 1.6 volts, and the valley current, I_v, is approximately 8 milliamperes.

The saturation resistance, r_s, can be found from the slope of the emitter characteristic in the saturation region (above 8 milliamperes) and is approximately 5.0 ohms.

The *diode voltage drop*, V_D, is defined as the forward voltage drop of the emitter junction. Since V_D is essentially equivalent to the forward voltage drop of a silicon diode, the value of V_D is dependent both upon forward current and temperature. V_D is particularly important since it is one of the major factors in peak voltage, V_p, of a UJT. The equation for V_p is:

$$V_p = V_D + \eta V_{B_2B_1} \tag{6-4}$$

In applications such as timers and oscillators, any changes in V_p would result in inaccuracy, since oscillator and timer circuit accuracy would be dependent upon the repeatability of V_p.

There are several ways to measure V_D. However, it is important to hold the emitter current near peak I_p when the measurement is made since it really is V_D at I_p that is required in equation 6-4. One simple way to measure V_D is shown in Fig. 6-10. A constant current signal equal to I_p is applied between the emitter and base one, and a potentiometric voltmeter is used to measure the voltage from emitter to base two. A potentiometric voltmeter has essentially infinite input impedance when the meter is nulled, and there is no current flowing in the base two circuit. Therefore, the voltage measured is equal to V_D.

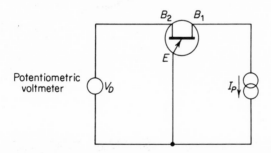

Fig. 6-10. Basic diode voltage drop V_D test circuit *(courtesy Motorola).*

Figure 6-11 shows V_D as a function of temperature for an emitter current of 1 microampere. The variation of V_D is essentially linear over the temperature range considered, and is equal to -2.7 mV/°C. The diode voltage drop therefore decreases with increasing temperature, and V_D can be written as:

$$V_D = V_{DN} - (T - 25)° K_D \tag{6-5}$$

where:

V_{DN} is the value of V_D at $T_A = 25°C$

K_D is -2.7 mV/°C. K_D is current dependent and the value for K_D given applies only at 1 microampere.

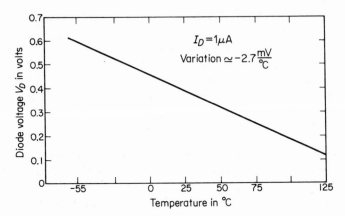

Fig. 6-11. Diode M voltage V_D versus temperature for annular UJT (*courtesy Motorola*).

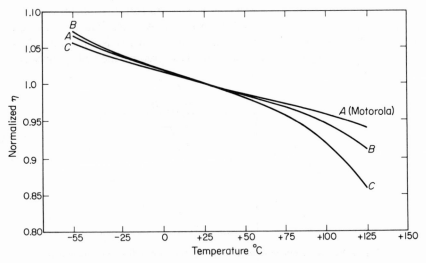

Fig. 6-12. Typical variations of standoff ratios versus temperature (*courtesy Motorola*).

The *intrinsic standoff ratio* is defined by:

$$\eta = \frac{V_p - V_D}{V_{B_2B_1}} = \frac{r_{B_1}}{r_{BB}} \qquad (6\text{-}6)$$

The intrinsic standoff ratio is somewhat temperature dependent. Figure 6-12 shows typical variations of standoff ratio with temperature for UJTs from three different manufacturers.

The intrinsic standoff ratio can be calculated if the $V_{B_2B_1}$, V_p, and V_D are known, using the equation 6-6. The ratio can also be measured directly with a sensitive ohmmeter (base one to emitter resistance divided by interbase resistance), using the equation 6-6. As in the case of any transistor, the voltages and currents used to make the resistance measurements should be kept well within the limits of the UJT.

The intrinsic standoff ratio is also slightly dependent upon $V_{B_2B_1}$, but the variation is so small that it can be ignored for practical purposes.

The *interbase resistance* r_{BB} is highly temperature dependent. A typical r_{BB} versus temperature characteristic curve for $V_{B_2B_1} = 3$ volts is shown in Fig. 6-13.

Interbase resistance r_{BB} is also found to vary with interbase voltage $V_{B_2B_1}$. A typical curve is shown in Fig. 6-14.

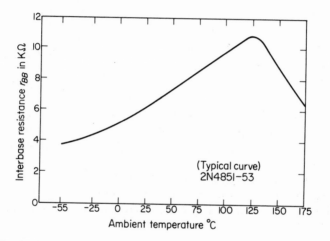

Fig. 6-13. Interbase resistance r_{BB} versus temperature (*courtesy Motorola*).

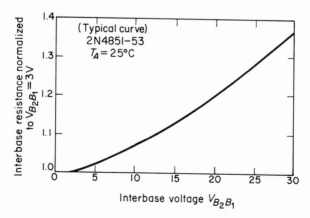

Fig. 6-14. Typical variation of r_{BB} as a function of interbase voltage $V_{B_2B_1}$ (*courtesy Motorola*).

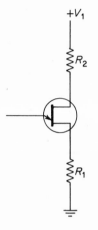

Fig. 6-15. Typical UJT bias circuit.

The *peak point characteristics* V_p and I_p decrease as temperature is increased. The variation of V_p with temperature can be minimized by the addition of an external resistor in the base two circuit, as shown in Fig. 6-15. With such a circuit (as the ambient temperature increases) the interbase resistance r_{BB} will increase and the interbase voltage $V_{B_2B_1}$ will also increase due to the voltage divider action of resistor R_2 (Fig. 6-15), r_{BB}, and R_1.

If R_2 is chosen correctly, the increase in interbase voltage will compensate for the decrease in V_D. The *approximate* value of R_2 is:

$$R_2 \cong \frac{0.70 r_{BB}}{\eta V_1} + \frac{(1-\eta)R_1}{\eta} \tag{6-7}$$

If R_2 satisfies Equation 6-7, the peak point voltage V_p will be given by:

$$V_p = \eta V_1 \tag{6-8}$$

The circuit of Fig. 6-15 is most effective if the temperature compensating resistor R_2 is in close physical contact with the emitter diode. This will compensate for changes in power dissipation, or sudden changes in ambient temperature.

The *valley point characteristics* V_v and I_V decrease as ambient temperature is increased. A curve showing typical temperature behavior for an annual (Motorola) UJT is shown in Fig. 6-16, where the curves are normalized to the value of 25°C.

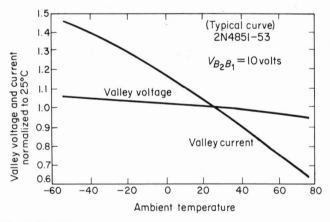

Fig. 6-16. Valley voltage and valley current versus ambient temperature (*courtesy Motorola*).

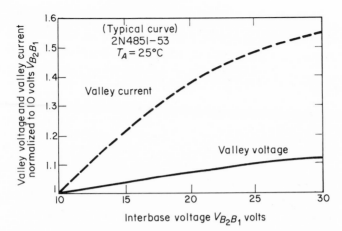

Fig. 6-17. Valley voltage and valley current versus interbase voltage $V_{B_2B_1}$
(*courtesy Motorola*).

V_v and I_V are also dependent upon the interbase voltage. When $V_{B_2B_1}$ increases, both V_v and I_V will increase also. Typical curves showing this behavior are given in Fig. 6-17, where the curves are normalized to the value of 10 volts $V_{B_2B_1}$.

Valley current I_V is a relatively difficult characteristic to measure since, as can be seen in Fig. 6-9, the voltage is relatively constant for large variations in emitter current around the valley point.

The *emitter reverse current* I_{EO} is generally specified as the current flowing from base two to the emitter with 30 volts applied between base two and the emitter (positive at base two), and the base one terminal open circuited.

I_{EO} is highly temperature dependent since it is (in effect) the leakage current of a silicon diode. A curve showing typical variation of I_{EO} with temperature for a 2N4851 is shown in Fig. 6-18 with I_{EO} being approximately 1.5 nanoampere at 25°C.

Interbase characteristics are usually indicated by measurement of base two current I_{B_2} as a function of interbase voltage $V_{B_2B_1}$ and emitter current I_E. Usually, interbase characteristics are measured on a *sweep basis*, rather than with constant voltages and currents, to avoid heating effects due to power dissipation.

A circuit for sweep test of interbase characteristics is shown in Fig. 6-19. A constant current is applied to the emitter from time t_0 to time t_1. Simultaneously, a voltage ramp going from 0 to 30 volts is applied to base two. Base one is grounded to complete the circuit. The current I_{B_2} is measured with a *current probe*, and applied to the vertical input of an oscilloscope. The voltage ramp, applied to the base two, is also applied to the oscilloscope horizontal input.

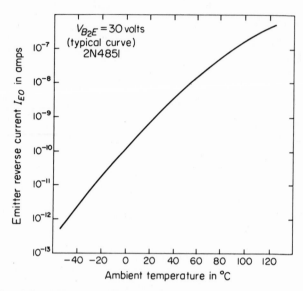

Fig. 6-18. Emitter reverse current I_{EO} versus ambient temperature (*courtesy Motorola*).

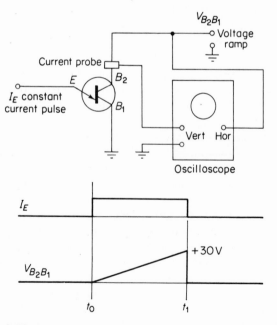

Fig. 6-19. Interbase characteristic test circuit (*courtesy Motorola*).

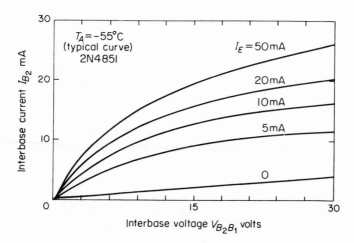

Fig. 6-20. Interbase characteristics at −55°C for 2N4851 (*courtesy Motorola*).

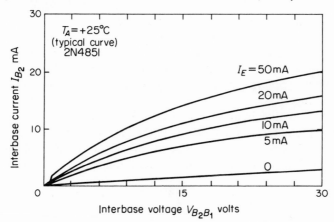

Fig. 6-21. Interbase characteristics at +25°C for 2N4851 (*courtesy Motorola*).

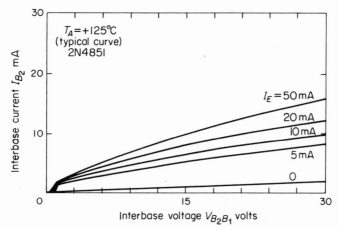

Fig. 6-22. Interbase characteristics at +125°C for 2N4851 (*courtesy Motorola*).

Figures 6-20, 6-21, and 6-22 show the interbase characteristics of a 2N4851 at ambient temperatures of $-55°C$, $+25°C$, and $+125°C$, respectively. As shown, the percentage increase in I_{B_2} decreases with increasing emitter current and temperature.

The *modulated interbase current* $I_{B_2(\text{mod})}$ is usually measured at $I_E = 50$ milliamperes, and $V_{B_2B_1} = 10$ volts. As shown in Fig. 6-21, the approximate modulated interbase current of a 2N4851 is 11 milliamperes.

The *transient characteristics* of a UJT are usually not specified in the same way as for a conventional junction transistor. For example, switching times are usually not specified on a UJT data sheet. Instead, a parameter of f_{max} is given which indicates the maximum frequency of oscillation that can be obtained using the UJT in a specified relaxation oscillator circuit.

However, in some applications such as critical timers, it may be of interest to determine *turn-on* and *turn-off* times associated with the UJT. The following describes procedures for these measurements.

The circuit of Fig. 6-23 can be used to measure t_{on} and t_{off} for the case where the UJT emitter circuit is *purely resistive*. Typical switching time values for a 2N4851 are: $t_{\text{on}} = 1$ microsecond; $t_{\text{off}} = 2.5$ microseconds. The waveform observed at the base one terminal when the UJT turns off is shown in Fig. 6-24.

Operation of the circuit in Fig. 6-23 is as follows: When the emitter is returned to ground, the stored charge in the junction will cause a current to flow out of the emitter, and the output voltage across R_1 will be smaller than the steady-state off value. Immediately following the removal of the

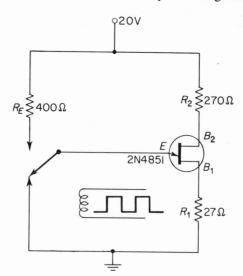

Fig. 6-23. Control circuit for turn-on and turn-off measurements (*courtesy Motorola*).

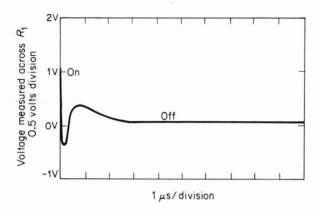

Fig. 6-24. "Turn-off" waveform for typical UJT *(courtesy Motorola).*

excess charge, the voltage across R_1 will go higher than the steady-state off value because r_{B_1} has still not returned to normal, following the conductivity modulation in the on state, and I_{B_2} will be larger than the steady-state off value.

The circuit of Fig. 6-25 can be used to measure t_{on} and t_{off} for the case where the UJT emitter circuit has *both capacitance and resistance* (as is usually the case). The test circuit of Fig. 6-25 is a *relaxation oscillator,* with turn-on and turn-off time being measured at the base one terminal. The turn-on and turn-off waveforms are shown in Figs. 6-26 and 6-27 respectively.

Turn-on time is measured from the start of the turn-on to the 90 per cent point. A typical turn-on time is 0.5 microsecond with the capacitance of C_E shown in Fig. 6-25. An increase in C_E capacitance will cause an increase in turn-on time. Turn-off time is measured from the start of turn-off to the 90 per cent point, and is about 12 microseconds (due to the long discharge

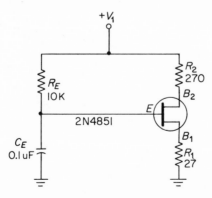

Fig. 6-25. Relaxation oscillator circuit for turn-on and turn-off tests.

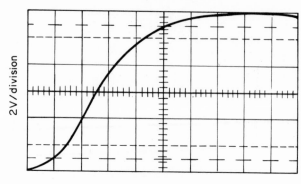

0.1 μsec/division

Fig. 6-26. "Turn-on" waveform for UJT using relaxation oscillator test circuit (*courtesy Motorola*).

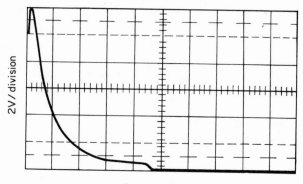

5 μ sec/division

Fig. 6-27. "Turn-off" waveform for UJT using relaxation oscillator test circuit (*courtesy Motorola*).

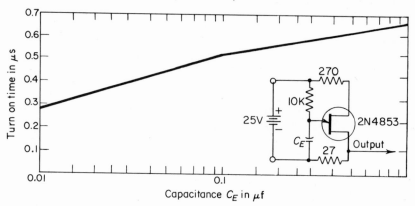

Fig. 6-28. Turn-on time versus emitter capacitance C_E (*courtesy Motorola*).

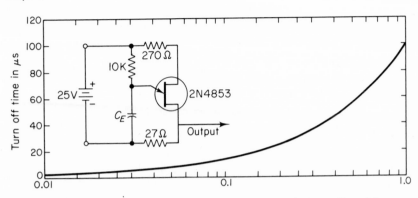

Fig. 6-29. Turn-off time versus emitter capacitance C_E (courtesy Motorola).

time of the capacitor). Turn-off time also increases with an increase in C_E capacitance.

The effect of C_E capacitance on switching time is shown in Figs. 6-28 and 6-29.

6-4. Comparison of Unijunction Characteristics

Table 6-2 lists typical characteristics for the three types of UJT structures discussed in Sec. 6-2.

TABLE 6-2

Comparison of Unijunction Characteristics

Characteristic	Structure Type (Typical Values)		
	Bar	Cube	Annular
Intrinsic Standoff Ratio η	0.6	0.65	0.7
Interbase Resistance r_{BB}	7 K	7 K	7 K
Emitter Saturation Voltage $V_{EB_1(\text{sat})}$	3 V	1.5 V	2.5 V
Peak-Point Current I_P	2 μA	1 μA	0.1 μA
Valley-Point Current I_V	15 mA	10 mA	7 mA
Emitter-Reverse Current I_{EO}	1 μA	0.1 μA	5 nA

6-5. Testing Unijunction Transistors

As in the case of conventional junction transistors, there are a number of commercial test sets for UJTs. Likewise, it is possible to adapt commercial test sets (designed to test junction transistors) for use with UJTs.

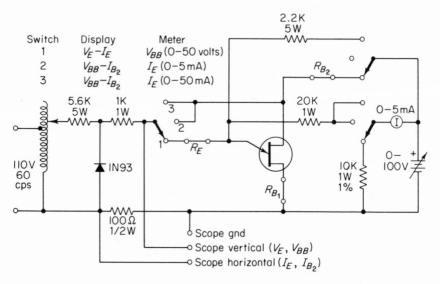

Fig. 6-30. Emitter and interbase curve tracer (*courtesy General Electric*).

The following describes circuits and procedures for testing of all major UJT characteristics.

Emitter and Interbase Curve Tracer. The circuit of Fig. 6-30 can be used to display either the emitter characteristic curves, or the interbase characteristic curves, on a standard oscilloscope. When the switches are set to display the interbase curves, the meter will indicate the emitter current. When the switches are set to display the emitter curves, the meter will indicate the interbase voltage.

It is important to set the variac and the d-c supply to zero *before* changing the switches, or inserting the UJT in the circuit, so as to avoid accidental burnout.

If desired, external resistors may be inserted in series with the emitter, base one, or base two, to determine the characteristics of the UJT in a particular circuit.

Commercial test equipment such as the Tektronix Transistor Curve Tracer can also be used to display the characteristic curves of the UJT. Connections which can be used with the Tektronix tracer are shown in Fig. 6-31. In the display of the emitter characteristic curves, the interbase voltage cannot exceed 12 volts because of the voltage limitation of the base current step generator.

General Purpose Test Set. A general purpose test set circuit for the UJT is shown in Fig. 6-32. This circuit can be used for testing of six major UJT characteristics as follows:

Test	Emitter curves	Interbase curves
Circuit	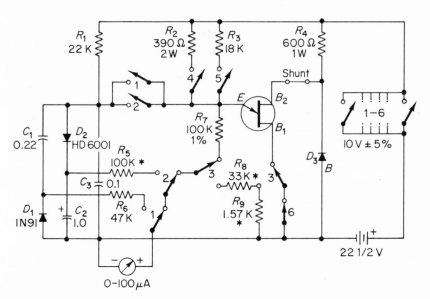	
Collector sweep polarity	+	+
Base step polarity	+	+
Collector peak volts range	200 V	20 V
Collector limiting resistor	5 K	500 Ω
Base current steps selector	20 mA/step	10 mA/step
Number of current steps	5	5
Vertical current range	2 mA/div I_E	2 mA/div I_{B_2}
Horizontal voltage range	1V/div V_E	2V/div V_{BB}

Fig. 6-31. Use of Tektronix 575 transistor curve tracer for display of UJT characteristic curves.

Note : (1.) All switches shown in the inactive position.

(2) All components ± 5% except as indicated.

(3) ✳ denotes approximate values, must be calibrated.

Fig. 6-32. General Electric UJT test set schematic (*courtesy General Electric*).

Test	Parameter	Range	Conditions
1	Peak-to-peak emitter voltage	0—10V	$V_{B_2B_1} = 10V$
			$C = 0.1$ microfarad
2	Intrinsic standoff ratio	0—1.0	$V_{B_2B_1} = 10V$
3	Interbase resistance	3K to infinity	Power < 15 milliwatts
4	Emitter saturation voltage	0—10V	$V_{B_2B_1} = 10V$
			$I_E = 50$ mA
5	Emitter voltage at 1 mA	0—10V	$V_{B_2B_1} = 10V$
			$I_E = 1$ mA
6	Emitter leakage current	0—100 micro-amperes	$V_{EB_2} \leq 10V$

The oscillator test portion of the circuit (Fig. 6-32) indicates if a UJT can oscillate in a relaxation oscillator circuit. Failure to pass this test usually means total failure of a UJT.

The intrinsic standoff ratio test makes use of a relaxation oscillator and a peak voltage detector. To calibrate the circuit for this test, a connection should be made from the common point of R_4-D_3 to the common point of D_2-C_2, and R_5 should be adjusted to give a full scale deflection on the meter with the button for test 2 depressed.

For measurement of interbase resistance, resistors R_8 and R_9 should be calibrated to give the correct meter deflection.

As an additional test, the interbase modulated current $I_{B_2(mod)}$ can be made by connecting an external meter in the base two shunt position, and depressing the button for test 4.

Intrinsic Standoff Ratio Test. The general purpose test circuit of Fig. 6-32 is usually satisfactory for measurement of the intrinsic standoff ratio. However, if greater accuracy is required, or if the equivalent emitter-diode voltage V_D is to be measured, the circuit of Fig. 6-33 can be used.

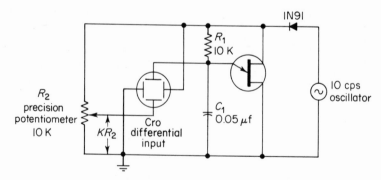

Fig. 6-33. Circuit for measurement of intrinsic standoff ratio and equivalent emitter diode voltage V_D (courtesy General Electric).

In the circuit of Fig. 6-33 the interbase voltage is swept by the 10 Hz oscillator, while the UJT oscillates at about 2 kHz in the simple relation oscillator. The interbase voltage is applied to the horizontal axis of the oscilloscope. The emitter voltage is applied to one input of the vertical differential amplifier, and the potentiometer voltage is applied to the other input.

The oscilloscope pattern consists of a plot of $V_{B_2B_1}$ on the horizontal axis against

$$V_P - KV_{B_2B_1} = V_D + (-K) V_{B_2B_1} \qquad (6\text{-}9)$$

on the vertical axis. K is the fractional setting of the potentiometer, and V_P is the upper envelope of the emitter voltage.

If K is set equal to the intrinsic standoff ratio, the upper envelope will be horizontal, and the displacement from the zero axis on the oscilloscope will be equal to V_D.

If a precision potentiometer is used for R_2, the intrinsic standoff ratio can be measured with an accuracy of better than 0.05 per cent, and V_D can be measured within 20 millivolts.

Peak-Point Current Test. The peak-point current I_P can be measured with the circuit of Fig. 6-34. This circuit measures the *minimum emitter current* required for oscillation in a relaxation oscillator circuit. This minimum emitter current is a good approximation to the peak-point emitter current. To measure I_P, the desired value of $V_{B_2B_1}$ is set and the voltage V_1 is increased until the UJT just fires, as indicated by a noise on the loudspeaker. The peak current is read on the microammeter. Care must be taken to avoid any ripple on the $V_{B_2B_1}$ supply since this can reduce the apparent peak-point current considerably.

Valley-Point Current and Voltage Test. The valley point corresponds to the point on a particular UJT emitter characteristic curve where the emitter voltage is at minimum, or where the dynamic resistance is zero. Because

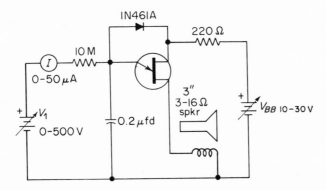

Fig. 6-34. Test circuit for measurement of peak-point emitter current I_p (courtesy Motorola).

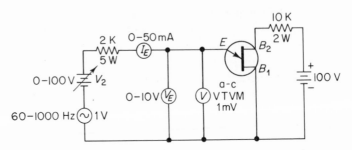

Fig. 6-35. Test circuit for measurement of valley voltage and valley current *(courtesy General Electric).*

of the slight change of the emitter voltage near the valley point, it is difficult to locate the position of the valley point by inspection of the emitter characteristic curve. The circuit of Fig. 6-35 can be used to measure V_v and I_V by the null method.

With the circuit of Fig. 6-35, the supply voltage V_2 is adjusted to obtain a null on the VTVM. The valley voltage and valley current can then be measured on the V_E and I_E meters, respectively.

The circuit of Fig. 6-35 shows typical bias conditions used for specification of I_V and V_v. Other values of interbase voltage and base two series resistance can be used as desired for special tests.

Frequency Response. The small-signal frequency-response characteristic of most value is the *emitter input-impedance* which can be measured with the circuit shown in Fig. 6-36. In this circuit, the UJT is biased in the negative resistance region, and a null is obtained on the VTVM by adjusting R_1 and C_1. At null the emitter input-impedance is equal to:

$$Z_E = -R_1 + \frac{j}{6.28C_1}$$

The capacitor C_2 is used to reduce the effective capacitance between the

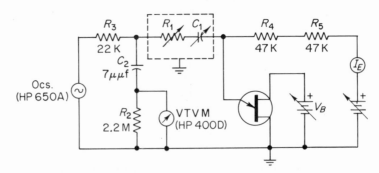

Fig. 6-36. Circuit for measurement of emitter input impedance of UJT *(courtesy General Electric).*

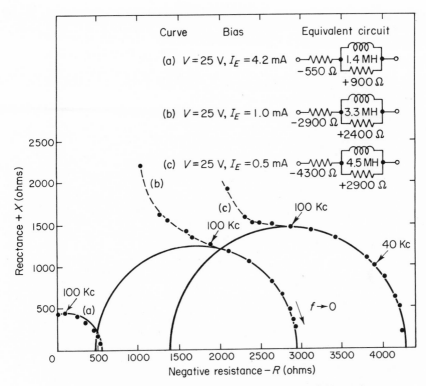

Fig. 6-37. Emitter impedance diagram of UJT at several bias points (*courtesy General Electric*).

emitter of the UJT and ground so as to prevent oscillation in the relaxation mode.

An impedance diagram for a typical UJT is shown in Fig. 6-37 for several bias conditions. Up to 100 kHz, the simple three-element RL network of Fig. 6-37 provides an accurate representation of the input impedance.

The frequency at which the resistive component of the emitter input-impedance is equal to zero is called the *resistance cutoff frequency* f_{RO}. This is a frequency figure of merit for a UJT since it represents the maximum frequency at which the UJT is regenerative at a given bias point.

The circuit of Fig. 6-36 can be used to measure f_{RO} if R_1 is set equal to zero, and a null obtained on the VTVM by adjusting the frequency and C_1. Since f_{RO} varies with bias conditions, measured values can be best presented by means of a plot such as that of Fig. 6-38 which gives contours of constant f_{RO} superimposed on the normal emitter characteristic curves. f_{RO} is proportional to the magnitude of the negative resistance at a given bias point. Therefore, a single measurement of f_{RO} will establish the values of f_{RO} over the entire negative resistance portion of the emitter characteristics.

The value of f_{RO} for a given emitter characteristic curve increases as the

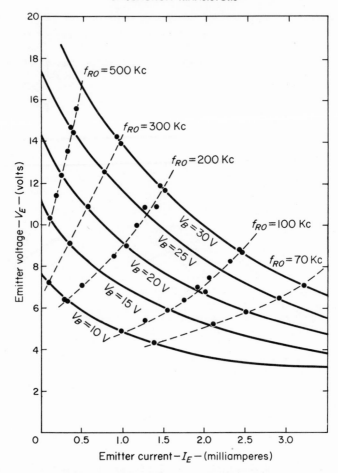

Fig. 6-38. UJT emitter characteristic curves with contours of constant resistance cutoff frequency (*courtesy General Electric*).

peak point is approached. The limiting value at the peak point corresponds to the *maximum frequency of oscillation* f_{max}. This is the maximum frequency at which the UJT can oscillate in a relaxation oscillator circuit.

For the lower range of $V_{B_2 B_1}$, where the heating caused by the interbase power dissipation is not excessive, f_{max} is directly proportional to $V_{B_2 B_1}$.

CHAPTER 7

Field Effect Transistors

The field effect transistor (or FET) is sometimes known as the semi-conductor that acts like a vacuum tube. This is because the characteristics of a FET are similar, but definitely not identical, to those of a pentode vacuum tube. There are two basic types of field effect transistors, known as the Junction Field Effect Transistor (or JEFT) and the Insulated Gate Field Effect Transistor (or IGFET). The IGFET is also known as the "metal oxide semiconductor" (or MOS), due to its physical structure.

The principles on which the JFET and IGFET operate (current controlled by an electric field) are very similar. The primary difference between the two is in the method by which the control element is made. This difference, however, results in a considerable difference in device characteristics.

7-1. Junction Field Effect Transistors (JFET)

The basic JFET is essentially a bar of doped silicon that acts like a re-sistor, see Fig. 7-1(a). The terminal into which current is injected is called the *source*. The source terminal is similar in function to the cathode of a vacuum tube. The opposite terminal is called the *drain* terminal, and can be likened to a vacuum tube plate. However, in a field effect transistor, the polarity of the voltage applied to the drain and source can be interchanged.

Current flow between source and drain is related to the drain-source voltage by the resistance of the material between the two terminals. In Fig. 7-1(b), P-type regions have been diffused into the N-type substrate. This leaves

113

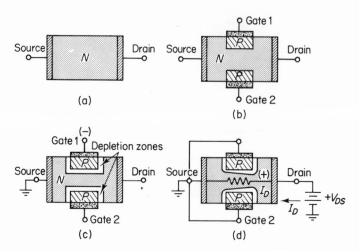

Fig. 7-1. Development of junction FETs (*courtesy Motorola*).

an *N*-type channel between the source and drain. (It is also possible to make a complementary JFET with a *P*-type channel by reversing all of the material types.) In the configuration of Fig. 7-1, the *P*-type regions are used to control the current flow between the source and the drain, and are thus called *gate regions*.

A depletion region surrounds the *PN* junctions when the junctions are reverse biased, as shown in Fig. 7-1(c). When the reverse voltage is increased, the depletion regions spread into the channel until they meet. This creates an almost infinite resistance between the source and drain.

Assume that a JFET is connected as shown in Fig. 7-1(d). The gate voltage (corresponding to a vacuum tube grid-bias voltage) is zero (grounded). Current flows from source to drain because of an external voltage V_{DS} (or drain-source voltage). This voltage corresponds to the $B+$ voltage of a vacuum tube.

A reverse bias is set up along the surface of the gate, parallel to the channel, because of the drain current flow in the channel. As the V_{DS} increases, the depletion regions again spread into the channel because of the voltage drop in the channel which reverse biases the junctions. The depletions will continue to grow with each increase in V_{DS} until the depletion regions meet. When this occurs, any further increase in V_{DS} is offset by an increase in the depletion region. In effect, there is an increase in channel resistance that prevents any further increase in drain current. The drain-source voltage that causes this current limiting condition is called the *pinch-off voltage* (or V_P). Further increases in drain-source voltage produce only a slight increase in drain current.

The symbols for a *P*-channel JFET and an *N*-channel JFET are shown in Fig. 7-2.

The drain-current (or I_D) versus drain-source voltage V_{DS}, with zero

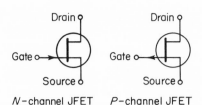

N−channel JFET   P−channel JFET

Fig. 7-2. P-channel and N-channel JFET symbols *(courtesy Motorola)*.

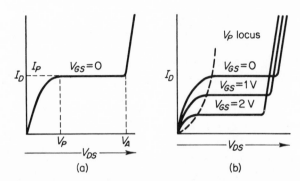

Fig. 7-3. Drain current characteristics *(courtesy Motorola)*.

gate-source voltage (or V_{GS}) is shown in Fig. 7-3(a). In the low-current region, the drain-current is linearly related to V_{DS}. As the I_D increases, the "channel" begins to deplete, and the slope of the I_D curve decreases. When the V_{DS} equals the pinch-off voltage V_P, the I_D reaches saturation, and stays relatively constant until drain-to-gain *avalanche* voltage (or V_A) is reached.

If a reverse voltage (corresponding to a vacuum tube negative grid bias) is applied to the gates, channel pinch-off occurs at a lower I_D as shown in Fig. 7-3(b). This is because the depletion region spread caused by the reverse-biased gates adds to that produced by V_{DS}, thus reducing the maximum current for any value of V_{DS}.

The geometry of a typical JFET is single-ended, as shown in Fig. 7-4.

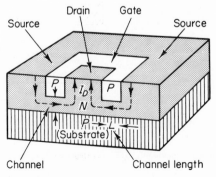

Fig. 7-4. Typical JFET construction *(courtesy Motorola)*.

This is because of the difficulty in diffusing impurities into both sides of a semiconductor wafer. In Fig. 7-4, the substrate is of *P*-type material onto which an *N*-type channel is grown. A *P*-type gate is then diffused into the *N*-type epitaxial channel.

The substrate, which functions as "Gate 2" of Fig. 7-1, is of relatively low resistance material to obtain maximum gain. For the same purpose, Gate 1 is of very low resistance material, allowing the depletion region to spread mostly into the *N*-type channel.

7-2. Insulated Gate Field Effect Transistors (IGFET)

The insulated gate field effect transistor operates with a slightly different control mechanism than the JFET. Figure 7-5 shows the development of an *N*-channel IGFET. The substrate is a high resistance *P*-type material. Two separate low resistance *N*-type regions (source and drain) are diffused into the substrate as shown in Fig. 7-5(b). The surface of the structure is covered with an insulating oxide layer as shown in Fig. 7-5(c). Holes are cut into the oxide layer, allowing metallic contact to the source and drain. The gate metal area is overlayed on the oxide, covering the entire channel region. Simultaneously, metal contacts to the drain and source are made as shown in Fig. 7-5(d). The contact to the metal area covering the channel is the gate terminal. There is no physical penetration of the metal through the oxide into the substrate. Since the drain and source are isolated by the substrate, any drain-to-source current in the absence of gate voltage is very low.

The metal area of the gate, in conjunction with the insulating oxide layer and the semiconductor channel, forms a capacitor. The metal area is the top plate; the substrate material is the bottom plate.

Operation of the gate is as follows (see Fig. 7-6): Positive charges at the metal side of the metal-oxide capacitor induce a corresponding negative

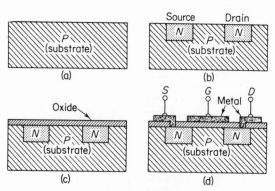

Fig. 7-5. Typical N-channel IGFET construction (*courtesy Motorola*).

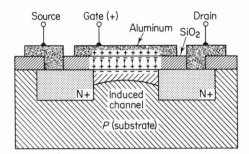

Fig. 7-6. Operation of typical IGFET (channel enhancement) (*courtesy Motorola*).

charge at the semiconductor side. As the positive charge at the gate is increased, the negative charge "induced" in the semiconductor increases until the region beneath the oxide becomes an *N*-type semiconductor region, and current can flow between source and drain through the "induced" channel. Drain current flow is *enhanced* by the gate voltage, and can therefore be modulated by the gate voltage. Channel resistance is directly related

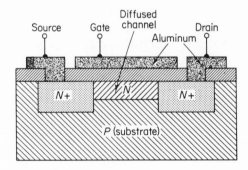

Fig. 7-7. Typical depletion mode IGFET construction (*courtesy Motorola*).

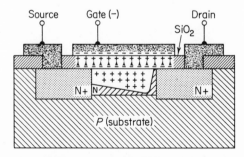

Fig. 7-8. Operation of typical IGFET (channel depletion) (*courtesy Motorola*).

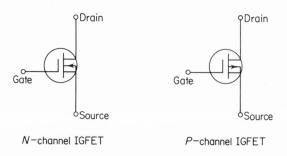

Fig. 7-9. P-channel and N-channel IGFET symbols (*courtesy Motorola*).

to the gate voltage. (It is also possible to make a complementary IGFET with a *P*-type channel by reversing all of the material types.)

The FET just described is called an *enhancement* type IGFET. A *depletion* type IGFET is formed when an *N*-type channel is diffused between the source and drain so that drain current can flow when the gate voltage is zero. Such an arrangement is shown in Figs. 7-7 and 7-8. The structure shown in Fig. 7-7 is both an enhancement and a depletion IGFET. When positive gate voltages are applied, the structure enhances in the same manner as that shown in Figs. 7-5 and 7-6. When negative gate voltages are applied, the channel begins to deplete of carriers as shown in Fig. 7-8.

It should be noted that with both the JFET and the depletion IGFET, drain current flow depletes the channel area nearest the drain terminal first.

The symbols for a *P*-channel IGFET and an *N*-channel IGFET are shown in Fig. 7-9.

7-3. Operating Modes of Field Effect Transistors

Field effect transistors can be operated in three modes: *depletion only*, *enhancement only*, and a combination of *enhancement and depletion*.

The basic differences between these three modes can most easily be understood by examining the transfer characteristics of Fig. 7-10.

The *depletion only* FET is classified as Type *A*, and has considerable drain current flow for zero gate voltage. No foward gate voltage is used. Drain current is reduced by applying a reverse voltage to the gate terminal. Most Type *A* devices are JFET.

The *depletion/enhancement* FET is classified as Type *B*, and also has considerable drain current flow for zero gate voltage (but not as much as Type *A*). Drain current is increased by application of a forward gate voltage, and reduced by application of a reverse gate voltage. Most Type *B* devices are IGFET. If a JFET is used as Type *B*, drain current can be increased by gate voltage *only* until the gate-source *PN* junction becomes forward biased.

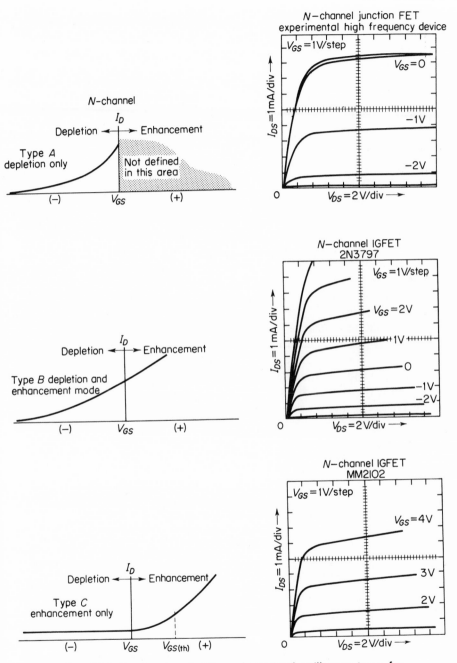

Fig. 7-10. Transfer characteristics and associated oscilloscope traces for Types A, B, and C FETs (*courtesy Motorola*).

At this point, a further increase in forward gate voltage will not produce an increase in drain current.

The *enhancement only* FET is classified as Type *C*, and has little or no current flow for zero gate voltage. Drain current does not occur until a forward gate voltage is applied. This voltage is known as the *threshold* voltage, and is indicated in Fig. 7-10 as V_{GS}(th). Once the threshold voltage is reached, the transfer characteristics for a Type *C* FET are similar to those of a Type *B*. Type *C* devices are always IGFET.

7-4. Equivalent Circuit of an IGFET

An equivalent circuit for the IGFET is shown in Fig. 11. Here, $C_{g(\text{ch})}$ is the distributed gate-to-channel capacitance, representing the oxide capacitance. G_{gs} is the gate-source capacitance of the metal gate area overlapping the source, while C_{gd} is the gate-drain capacitance of the metal gate area overlapping the drain. $C_{d(\text{sub})}$ and $C_{s(\text{sub})}$ are junction capacitances from drain to substrate, and source to substrate. Y_{fs} is the transadmittance between drain current and gate source voltage. The modulated channel resistance is r_{ds}. R_D and R_S are the bulk resistance of the drain and source.

The input resistance of the IGFET is exceptionally high because the gate behaves as a capacitor with a very low leakage. A typical input resistance is 10^{14} ohms. The output impedance is a function of r_{ds} (which is related to gate voltage) and the drain and source bulk resistances (R_D and R_S).

To turn the IGFET on, the gate-channel capacitance $C_{g(\text{ch})}$, the miller capacitance C_{gd}, and the drain-substrate capacitance $C_{d(\text{sub})}$ must be charged. The resistance of the substrate determines the peak charging current of $C_{d(\text{sub})}$.

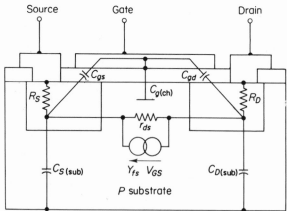

Fig. 7-11. Equivalent circuit of enhancement mode IGFET (*courtesy Motorola*).

7-5. Static Characteristics of Field Effect Transistors

The primary static (or d-c) characteristics of a FET are those that indicate the effect of a control (or gate) signal on the output current. Figure 7-12 lists the data sheet specifications normally used to describe the V_{GS}-I_D curves shown in Fig. 7-10. The data in Fig. 7-12 show the characteristics for the three basic operating modes of FET. Each of the characteristics is provided with a description, as well as test diagrams.

All of the characteristics shown in Fig. 7-12 are for FETs connected in the usual triode configuration. Sometimes, JFETs are connected as tetrodes. That is, the two gates are separately accessible for a control signal (or gate voltage). Under such conditions, an additional characteristic must be considered—the voltage that must be applied to one gate to produce cutoff when only the opposite gate is connected to the source. The gate voltage required for drain current cutoff with only one of the gates connected to the source is always higher than that for the usual triode-connected case where both gates are tied together.

When a JFET is connected as a tetrode, the gate 1 to source cutoff voltage (with gate 2 connected to the source) is identified as $V_{G_1S(\text{off})}$, while the gate 2 to source cutoff voltage (with gate 1 connected to the source) is identified as $V_{G_2S(\text{off})}$.

Another specification that applies only to a JFET connected as a tetrode is *reach-through* voltage. This is the difference voltage that may be applied to the two gates before the depletion region of one spreads into the junction of the other. Reach-through is an undesirable condition since it causes a decrease in input resistance as a result of an increased gate current. Large amounts of reach-through current can destroy a FET.

Gate leakage is a particularly important characteristic of a FET, since it is directly related to resistance. When gate leakage is high, input resistance is low, and vice versa. Gate leakage is usually specified as I_{GSS} (reverse-bias, gate-to-source current with the drain shorted to the source), and is tested with a circuit similar to that of Fig. 7-13.

Normally, input resistance of a FET is very high since the leakage current across a reverse-biased *PN* junction (in the case of a JFET) and across a capacitor (in the case of an IGFET) is very small. At a temperature of 25°C, the JFET input resistance is approximately several hundred megohms, while the IGFET input resistance is even greater. Input resistance may decrease by several orders of magnitude in a JFET as temperature is raised to 150°C, since leakage usually increases with temperature in *PN* junction devices. Therefore, JFETs usually have gate leakage specified at two temperatures. IGFETs are not drastically affected by temperature, and their input resistance remains extremely high, even at high temperatures.

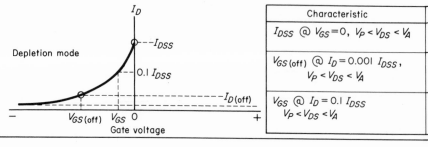

Characteristic
I_{DSS} @ $V_{GS}=0$, $V_P < V_{DS} < V_A$
$V_{GS(off)}$ @ $I_D = 0.001\ I_{DSS}$, $V_P < V_{DS} < V_A$
V_{GS} @ $I_D = 0.1\ I_{DSS}$ $V_P < V_{DS} < V_A$

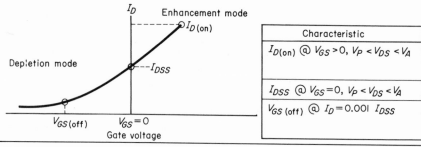

Characteristic
$I_{D(on)}$ @ $V_{GS} > 0$, $V_P < V_{DS} < V_A$
I_{DSS} @ $V_{GS}=0$, $V_P < V_{DS} < V_A$
$V_{GS(off)}$ @ $I_D = 0.001\ I_{DSS}$

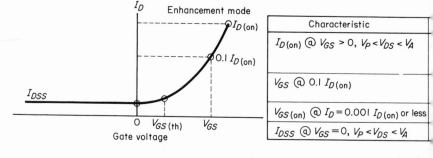

Characteristic
$I_{D(on)}$ @ $V_{GS} > 0$, $V_P < V_{DS} < V_A$
V_{GS} @ $0.1\ I_{D(on)}$
$V_{GS(on)}$ @ $I_D = 0.001\ I_{D(on)}$ or less
I_{DSS} @ $V_{GS}=0$, $V_P < V_{DS} < V_A$

Fig. 7-12. Static characteristics for

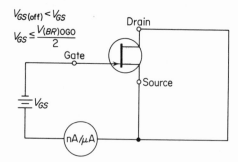

Fig. 7-13. Test circuit for leakage current (*courtesy Motorola*).

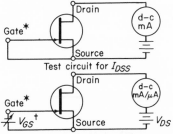

Description
Zero–gate–voltage drain current Represents maximum drain current
Gate voltage necessary to reduce I_D to some specified negligible value at the recommended V_{DS}, i.e. cutoff
Gate voltage for a specified value of I_D between I_{DSS} and I_{DS} at cutoff–normally 0.1 I_{DSS}

Test circuit for I_{DSS}

Test circuit for V_{GS} and $V_{GS\,(off)}$

* Gates internally connected
† Adjust for desired I_D

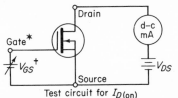

Description
An arbitrary current value (usually near max rated current) that locates a point in the enchancement operating mode
Zero–gate–voltage drain current
Voltage necessary to reduce I_D to some specified negligible value at the recomended V_{DS}, i.e. cutoff

Test circuit for $I_{D(on)}$

* Gates internally connected
† Adjust for desired I_D, normally near max-rated I_D

Test circuits for I_{DSS} and $V_{GSS\,(off)}$ same as type A

Description
An arbitrary current value (usually near max rated current) that locates a point in the enhancement operating mode
Gate-source voltage for a specified drain current of 0.1
Gate cutoff or turn-on voltage
Leakage drain current

$I_{D(on)}$ test circuit same as for type B

V_{GS} test circuit same as for $I_{D(on)}$

$V_{GS\,(th)}$ test circuit same as $V_{GS\,(off)}$ in type A, except reverse V_{GS} battery polarity

I_{DSS} test circuit same as for type A

three FET types (courtesy Motorola).

Some data sheets specify gate leakage current as I_{GDO} (leakage between gate and drain with the source open). Other data sheets use the term I_{GSO} (leakage between gate and source with the drain open). When these characteristics are used, the gate leakage current is lower than when I_{GSS} is specified. Consequently, I_{GDO} and I_{GSO} do not represent the "worst case" condition, making I_{GSS} the preferred specification.

Voltage breakdown is another particularly important characteristic of a FET. There are a number of specifications to indicate the maximum voltage that can be applied to the various elements of a FET. These include:

$V_{(BR)GSS}$ or gate-to-source breakdown voltage

$V_{(BR)DGO}$ or drain-to-gate breakdown voltage

$V_{(BR)DSX}$ or drain-to-source breakdown voltage (normally used only for IGFETs)

In some specification sheets, there may be ratings and specifications indicating the maximum voltages that may be applied between the individual gates and the drain and source (for tetrode-connected devices), between drain and gate, and so on. Not all of these specifications are found on every data sheet. Some of the specifications provide the same information in slightly different form. However, by understanding the various breakdown mechanisms, it should be possible to interpret the intent of each specification and rating.

For example, in JFETs, the maximum voltage that may be applied between any two terminals is the lowest voltage that will lead to breakdown of the gate junction. In the case of $V_{(BR)GSS}$ [shown in Fig. 7-14(a)], an increasingly higher reverse voltage is applied between the common gates and the source. Junction breakdown can be determined in gate current (beyond normal I_{GSS}) which indicates the beginning of an "avalanche" (V_A of Fig. 7-3) condition.

For a JFET, the specification $V_{(BR)DGO}$ provides the same information as $V_{(BR)GSS}$. In this case, the gates are connected together and an increasing voltage is applied between drain and gates. When the applied voltage becomes high enough, the drain-gate junction will go into avalanche, indicated by an increase in gate current (beyond I_{DGO}). For both specifications, breakdown should normally occur at the same voltage value.

In FET specifications, "breakdown" voltage is interchangeable with the term "avalanche" voltage (V_A).

Avalanche occurs at a lower value of V_{DS} when the gate is reverse biased than for the zero-bias condition. This condition is shown in Fig. 7-3, and is caused by the fact that the reverse-bias gate-voltage adds to the drain voltage, thereby increasing the effective voltage across the junction. The maximum amount of drain-source voltage that may be applied is $V_{DS(max)} = V_{(BR)DGO} - V_{GS}$.

For IGFETs, the breakdown voltage is determined by a different set of

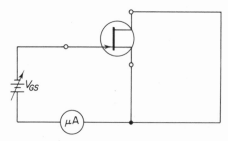

Fig. 7-14. Test circuit for breakdown voltage (courtesy Motorola).

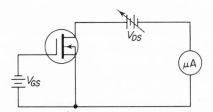

Fig. 7-15. Test circuit for drain-to-source breakdown of IGFETs (courtesy Motorola).

factors. Since the gate is insulated from the drain-source and channel by an oxide layer, the *gate breakdown voltage* is dependent upon the thickness and purity of the layer. Gate breakdown voltage is the voltage that will physically puncture the oxide layer.

Drain-to-source breakdown voltage for an IGFET must also be determined by different factors. For Type C IGFETs, with the gate connected to the source (in the cutoff condition) and the substrate floating, there is no effective channel between drain and source, and the applied drain-source voltage appears (in effect) across two back-to-back connected, reverse-biased diodes. These diodes are represented by the source-to-substrate and substrate-to-drain junctions. Drain current remains at a very low (picoampere) level as drain voltage is increased until drain voltage reaches a value that causes reverse (reach-through or punch-through) breakdown of the diodes. This condition, represented by $V_{(VR)DSS}$, is indicated by an increase in I_D above the I_{DSS} level, as shown in Fig. 7-15.

For Type B IGFETs, the $V_{(BR)DSS}$ symbol is sometimes replaced by $V_{(BR)DSX}$. The main difference between the two symbols is the replacement of the last subscript "S" with the subscript "X." The "S" normally indicates that the gate is shorted to the source; "X" indicates that the gate is biased to cutoff, or beyond. To obtain cutoff in Type B IGFETs, a depletion bias voltage must be applied to the gate, as shown in Fig. 7-15.

The *"on" drain-source voltage*, or $V_{DS(on)}$, is an important static characteristic for switching FETs. For an IGFET used in switching circuits, the $V_{DS(on)}$ is similar to the collector-emitter saturation-voltage base-current characteristics $V_{CE(sat)} - I_B$ of a conventional junction transistor.

7-6. Dynamic Characteristics of Field Effect Transistors

Unlike the static characteristics, the dynamic characteristics (a-c or signal) of FETs apply equally to Types A, B, and C. However, the conditions and presentations of the dynamic characteristics depend mostly on the intended *application*. Table 7-1 indicates the dynamic characteristics needed to describe a FET adequately for various applications.

TABLE 7-1

Typical FET Dynamic Characteristics

Audio	RF-IF	Switching	Chopper
Y_{fs}(1 kHz)	Y_{fs}(1 kHz)	Y_{fs}(1 kHz)	Y_{fs}(1 kHz)
C_{iss}	C_{iss}	C_{iss}	C_{iss}
C_{rss}	C_{rss}	C_{rss}	C_{rss}
Y_{os}(1 kHz)	Y_{fs}(HF)	$Cd_{(sub)}$	$Cd_{(sub)}$
Y_{fs}(HF)	Re(Y_{os})(HF)	$r_{ds(on)}$	$r_{ds(on)}$
NF	Re(Y_{os})(HF)	t_{d_1}, t_{d_2}	
	NF	t_r, t_f	

Forward transadmittance (or transconductance) Y_{fs} is the most important dynamic characteristic for FETs, no matter what application. Y_{fs} serves as a basic design parameter in audio and *RF*, and is a widely accepted device figure of merit.

Because FETs have many characteristics similar to those of vacuum tubes, the symbol g_m is sometimes used instead of Y_{fs}. This is further confused, since the g notation school also uses a number of subscripts. In addition to g_m, some data sheets show g_{fs}, while others go even further out with g_{21}.

No matter what symbol is used, Y_{fs} defines the relationship between an input signal *voltage* and an output signal *current*, with the drain-source voltage held constant, or

$$Y_{fs} = \Delta I_D / \Delta V_{GS}\Big|_{V_{DS}=K}$$

(7-1)

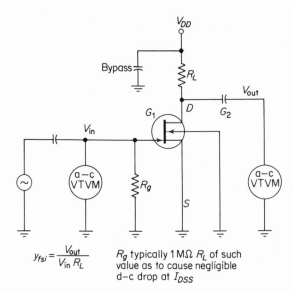

$$y_{fsi} = \frac{V_{out}}{V_{in}\, R_L}$$

R_g typically 1 MΩ R_L of such value as to cause negligible d–c drop at I_{DSS}

Fig. 7-16. Test circuit for Y_{fs} (courtesy Motorola).

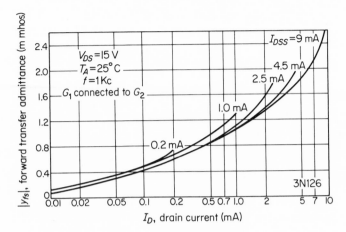

Fig. 7-17. Forward transfer admittance versus drain current for typical 3N126 JFETs (courtesy Motorola).

Y_{fs} is expressed in *mhos* (current divided by voltage). Figure 7-16 is a typical Y_{fs} test circuit for a tetrode-connected JFET.

In most data sheets, Y_{fs} is specified at 1 kHz, with a V_{DS} the same as that for which $I_{D(on)}$ is obtained.

Since the $I_D - V_{GS}$ curve of a FET is nonlinear, Y_{fs} will vary considerably with changes in I_D. This variation, for a typical N-channel Type A JFET (the 3N126) is illustrated in Fig. 7-17.

Three Y_{fs} measurements are usually specified for tetrode-connected FETs. One of these, with two gates tied together, provides a Y_{fs} value for the condition where a signal is applied to both gates simultaneously. The other two measurements provide the Y_{fs} for the two gates individually. Generally, with the two gates tied together, Y_{fs} is higher, and more gain may be realized in a given circuit. However, because of the increased capacitance, gain-bandwidth product is much lower.

For FETs used at radio frequencies, an additional value of Y_{fs} should be specified at or near the highest frequency of operation. This value should also be measured at the same voltage conditions as those used for $I_{D(on)}$.

Because of the importance of the imaginary component at radio frequencies, the high-frequency Y_{fs} specification should be a complex representation, and should be given either in the specification table or by means of curves showing typical variations, as in Fig. 7-18 for the 3N126 JFET.

The real portion of this high-frequency Y_{fs}, such as $Re(Y_{fs})(\text{HF})$ or G_{21}, is usually considered as a significant figure of merit.

Output admittance Y_{os} is another important dynamic characteristic for FETs. Y_{os} is also represented by various Y and g parameters, such as Y_{22}, g_{os}, and g_{22}. Y_{os} is even specified in terms of drain resistance, or r_d, where

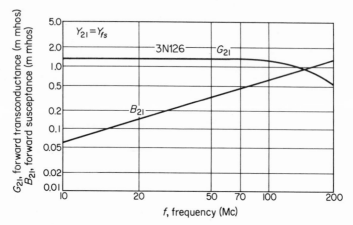

Fig. 7-18. Forward transfer admittance versus frequency
(courtesy Motorola).

$r_d = 1/Y_{os}$. This is similar to the vacuum tube characteristic of output admittance, where $Y_{output} = 1/r_p$, or output admittance is equal to one divided by plate resistance.

No matter what symbol is used, Y_{os} defines the relationship between output signal current and output voltage, with the input voltage held constant, or

$$Y_{os} = \Delta I_D/\Delta V_{DS}\Big|_{V_{GS}=K} \qquad (7\text{-}2)$$

Y_{os} is expressed in mhos (current divided by voltage). Figure 7-19 is a typical Y_{os} test circuit for Types A and B FETs, which are measured with the gates and source grounded. For Type C FETs, Y_{os} is measured at some specified value of V_{GS} which will permit a substantial drain current to flow.

Voltages and frequencies for measuring Y_{os} should be exactly the same as for Y_{fs}. Y_{os} is a complex number, and should be specified as a magnitude at 1 kHz, and in complex form at high frequencies.

Amplification factor, μ, does not usually appear on most FET specification

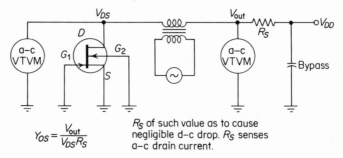

$$Y_{OS} = \frac{V_{out}}{V_{DS}R_S}$$

R_S of such value as to cause negligible d-c drop. R_S senses a-c drain current.

Fig. 7-19. Test circuit for Y_{os} (Types A and B FETs) (courtesy Motorola).

sheets. This is because amplification factor does not usually have a great significance in most small-signal applications. However, amplification factor is sometimes used as a figure of merit.

Amplification factor defines the relationship between output signal voltage and input signal voltage, with the output current held constant, or

$$\mu = \Delta V_{DS}/\Delta V_{GS}\Big|_{I_D=K} \tag{7-3}$$

Amplification factor can also be calculated by Y_{fs}/Y_{os}.

The *input capacitance with a common source circuit* C_{iss} is used on specification sheets for low-frequency FETs. In the specification sheets used for

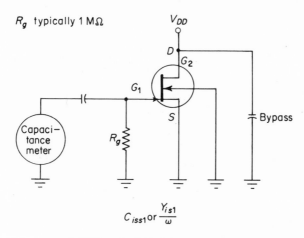

(a)With gate 2 tied to source

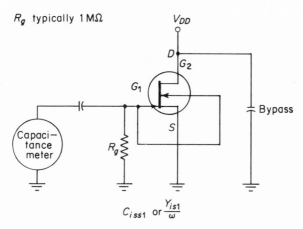

(b) With common gates

Fig. 7-20. Test circuit for C_{iss} in tetrode JFETs (*courtesy Motorola*).

FETs operated at radio frequencies, *input admittance with a common source circuit* Y_{is} is more popular. This is because Y_{is} is entirely capacitive at low frequencies.

Figure 7-20 shows typical circuits for test of C_{iss} in tetrode JFETs. As with Y_{fs}, two measurements (one with gate 2 tied to source, and the other with common gates) are necessary for tetrode JFETs.

At very high frequencies, the real component of Y_{is} becomes important so that RF FETs should have Y_{is} specified as a complex number at the same conditions as other high-frequency parameters.

For switching, application C_{iss} is of major importance since a large voltage swing at the gate must appear across the input capacitance C_{iss}.

Reverse transfer capacitance with a common source circuit C_{rss} is used on specification sheets for both low- and high-frequency FETs. *Reverse transfer admittance* Y_{rs} is rarely, if ever, used. This is because the Y_{rs} for a FET remains almost completely capacitive, and relatively of constant capacity, over the entire usable FET frequency spectrum. Consequently, the low-frequency C_{rss} is an adequate specification.

Figure 7-21 shows typical circuits for testing of C_{rss} in tetrode JFETs. Again, two measurements (gate 1 individually, and common gates) are necessary for tetrode JFETs.

For switching, application C_{rss} is also of major importance. Similar to the C_{ob} of a conventional junction transistor, C_{rss} must be charged and discharged during the switching interval.

For chopper applications C_{rss} is the feed-through capacitance for the chopper drive.

C_{rss} is also known as the Miller-effect capacitance, since the reverse

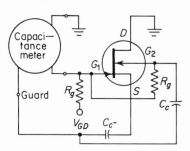
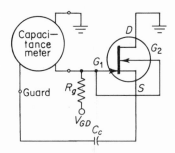

(a) $V_{GD} = -V_{DS}$
R_g typically 1 MΩ
$V_{G_1 G_2} = 0$ to prevent "reach through"
C_c ac source and G_2 to guard signal

(b) $V_{GD} = -V_{DS}$ $I_D = 0$
R_g typically 1 MΩ C_c couples guard signal
$V_{G_1 G_2} = 0$ to source terminal

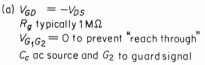

$$C_{rss} \text{ or } \frac{Y_{rs1}}{\omega}$$

$$C_{rss12} \text{ or } \frac{Y_{rs12}}{\omega}$$

Fig. 7-21. Test circuits for C_{rss} (courtesy Motorola).

capacitance can produce a condition similar to the Miller effect in vacuum tubes (such as constantly changing frequency response curves).

The *drain-substrate junction capacitance* $C_{d(sub)}$ is an important characteristic for IGFETs used in switching circuits. This is because $C_{d(sub)}$ appears in parallel with the load in a switching circuit, and must be charged and discharged between the two logic levels during the switching interval. No special circuit or conditions need be used to measure drain-substrate junction capacitance $C_{d(sub)}$.

Noise figure NF for FETs is usually specified as "spot noise." This refers to noise at a particular frequency. The noise figure will vary with frequency and also with the input resistance. Figure 7-22 is a typical noise figure graph for 3N126 FET. As shown, the drain voltage, drain current, and ambient temperature should be specified, in addition to the frequency and input resistance.

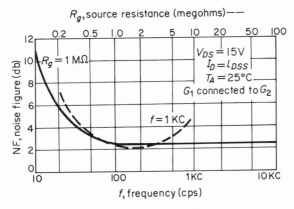

Fig. 7-22. Typical variations of FET noise figure with frequency and source resistance (*courtesy Motorola*).

Channel resistance $r_{ds(on)}$ describes the bulk resistance of the channel in series with the drain and source. Channel resistance is particularly important for switching and chopper circuits since it affects the switching speed, and determines the output level. Channel resistance is also described as $r_{d(on)}$, r_{DS}, and r_{ds}, depending upon the data sheet.

Figure 7-23 shows a typical circuit for testing of channel resistance in a JFET. Both gates should be tied together, and all terminals should be at the same (zero) d-c voltage. The a-c voltage should be low enough so that there is no pinch-off in the channel. IGFETs can be tested with essentially the same circuit. However, the gate should be biased so that the IGFET is operating in the enhancement mode.

Switching time characteristics are often specified on FET data sheets. These include t_{d_1} (delay time), t_{d_2} or t_{ds} (storage time), t_r (rise time), t_f (fall

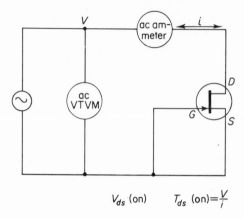

$$V_{ds}\ \text{(on)} \qquad T_{ds}\ \text{(on)} = \frac{V}{i}$$

Fig. 7-23. Test circuit for JFET channel resistance (*courtesy Motorola*).

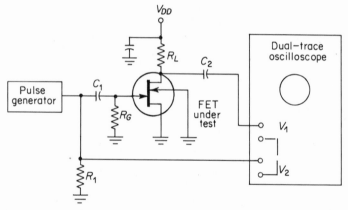

R_G =1 megohm (typical)
R_L = value that will cause negligible dc drop at I_{DSS}
R_1 =50 ohms (typical)

Fig. 7-24. Typical test circuit for measurement of FET switching
characteristics.

time). These characteristics must be measured with a pulse source, and a
multiple trace oscilloscope, as are similar characteristics for conventional
junction transistors.

Figure 7-24 shows typical circuits for switching-time tests of a FET.
The procedures are identical to those for conventional junction transistors,
described in Chapter 3.

CHAPTER 8

Basic Transistor Circuit Design

This chapter discusses the practical design aspects of transistor circuits. It is not intended that this publication be considered a design handbook. However, it is often of great value if the laboratory technician understands the problems of basic transistor circuit design.

The procedures of this chapter show the basic steps required to design an *RC*-coupled Class *A* transistor amplifier, using the information available on the transistor data sheet. Although there are many variations of this circuit, the basic design steps remain the same.

The design techniques described in this chapter (or similar techniques) have been in use for many years, and are often termed the *Current Design Method.*

8-1. Determining Design Characteristics

The first step in design is to determine the characteristics of the circuit and the transistor to be used. For example, will the amplifier be used in the audio frequency range, or will it be required to pass higher frequencies? What power-supply voltages will be available? What transistor is to be used?

In the following example, it is assumed that an amplifier using the basic circuit of Fig. 8-1 is to be designed, that a 2N332 *NPN* transistor is to be used, that the power supply voltage is 10 volts, and that the amplifier will be used with signals up to 100 kHz.

Once these preliminary factors have been decided, the following design

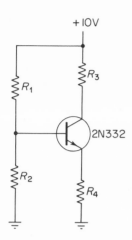

+ 10V

R_1

R_3

2N332

R_2

R_4

Fig. 8-1. Basic circuit used for Current Design Method.

data should be recorded. This will require reference to the transistor data sheet, as well as some arbitrary decisions.

Available (or desired) power-supply voltage. An arbitrary 10 volts has been selected. The voltage must be *below* the collector-emitter breakdown voltage (BV_{CBO}). The rated collector-emitter breakdown voltage for a 2N332 is 45 volts, so the arbitrary 10 volts is well within tolerance.

Maximum current rating. The maximum rated current for a 2N332 is 25 milliamperes, so this factor is taken directly from the data sheet.

Maximum power dissipation. The maximum rated power dissipation for a 2N332 is 150 milliwatts at 25°C, and 50 milliwatts at 125°C. When the exact operating temperature is not known, use from one-half to one-fourth of the 25°C rating. An arbitrary 50 milliwatts is chosen for the example.

Leakage current (I_{CBO}). The maximum rated leakage current for a 2N332 is 2 microamperes at 25°C, and 50 microamperes at 150°C. When the exact operating temperature is not known, use 10 times the 25°C rating. An arbitrary 20 microamperes is chosen for the example.

Total output capacitance. The total output capacitance of the circuit is determined by adding the transistor output capacitance, the input capacitance of the following circuit, and stray capacitance. While the output capacitance of the transistor can be taken from the data sheet, it is difficult to determine exact input capacitance of the following stage, and almost impossible to find stray capacitance. However, for a typical *RC*-coupled amplifier operating at frequencies up to 5 MHz, a value of 50 to 100 picofarads will usually be satisfactory. An arbitrary 50 picofarads is chosen for the example.

Maximum operating frequency. The rated cutoff frequency for a 2N332 is 15 MHz. The maximum operating frequency for the circuit should be approximately 1/10 of the rated cutoff frequency. In the case of the 2N332, this would be 1.5 MHz, which is well above the desired 100 kHz specified for the example.

8-2. Operating Load Current

Once the basic design characteristics have been tabulated, the next step is to calculate the operating load current.

Two major factors determine operating load current for a transistor amplifier: *leakage current* and *maximum rated current*.

Obviously, the load current can not exceed the maximum rated current for the transistor (25 mA). Likewise, the load current must not be less than the leakage current (20 microamperes), or no current will flow.

If the amplifier is to be operated from a battery, an operating load current near the low end should be selected to minimize battery drain. However, the load current should be no less than 10 times the leakage current. Since leakage current is 0.020 mA, the *minimum load current* should be 0.2 mA.

At the high end, the maximum load current should not exceed the maximum power dissipation with maximum voltage. Using the arbitrary 50 mW, divided by 10 volts ($I = P/E$), the maximum load current would be 5 mA.

The operating load current should be midway between these two extremes: $5 - 0.2 = 4.8$ divided by $2 = 2.4$ mA.

8-3. Operating Load Resistance

Once the operating load current has been established, the next step is to calculate the load resistance value.

When operating at frequencies up to about 100 kHz, the load resistance value can be calculated on a direct current basis, using basic Ohm's law. When the operating load current is flowing, the collector load resistor should drop the collector supply voltage to one-half.

In the example, $10 \times 0.5 = 5$ volts divided by 2.4 mA ($R = E/I$), the load resistance value should be 2083 ohms. A 2100-ohm collector resistor R_3 (or nearest stock value with a 5 per cent tolerance) should be satisfactory.

If the amplifier were to operate at frequencies considerably higher than 100 kHz, it might be necessary to reduce the collector load resistance to extend the high frequency response. The total output capacitance (an arbitrary 50 pF in the example) must charge and discharge through the collector load resistance. When the operating frequency is such that the reactance of the total output capacitance is equal to the load resistance, the output will drop about 3 dB. (If the load resistance is so high that the output capacitance never fully charges or discharges, the output will drop completely.) However, since the frequency at which a 3-dB drop occurs is usually considered at the top limit, the collector load resistance should be equal to (or lower than) the reactance of the output capacitance at the desired top frequency.

Assume that the amplifier were to be used at frequencies up to 5 MHz, such as an oscilloscope pre-amplifier. In this case, the reactance of the output capacitance would be: $X_c = 1/(6.28 \times 5^6 \times 50^{-12}) = 637$ ohms.

Then the collector load resistance should be 637 ohms (or less). The amplifier output will drop off approximately 3 dB at 5 MHz.

When in doubt as to the collector load resistance value, calculate the value both ways (on a d-c basis, and using the reactance of the output capacitance at the highest desired operating frequency). Then use the lowest of the two values.

8-4. Bias Resistance Values

Once a desired load current and load resistance have been selected, the next step is to calculate the bias resistance values that will produce the load current under no-signal conditions.

The value of the emitter resistance (R_4 in Fig. 8-1) should be approximately 1/5 that of the load resistance. Using a 2100-ohm load resistor R_3, the emitter resistance R_4 would be 425 ohms (or the nearest 5 per cent stock value). With a 425-ohm emitter resistor R_4, approximately 1/10 of the total power-supply voltage (or 1 volt) will be dropped across R_4. The emitter voltage will, therefore, be 1 volt, while the collector is at 5 volts (under no-signal conditions with 2.4 mA current).

To calculate the bias resistance values (R_1 and R_2) it is necessary to make two assumptions.

First, there will be some voltage drop across the forward-biased base-emitter junction. This junction drop will be approximately 0.2 volt for germanium transistors, and 0.5 volt for silicon transistors. Since the 2N332 is silicon, use the arbitrary 0.5-volt drop.

Second, assume that the current through R_1 and R_2 will be 1/10 of the load current: 2.4 mA $\times$ 0.1 = 0.24 mA.

The value of voltage at the junction of R_1 and R_2 (the base) should be 1/10 of the total power-supply voltage (or 1 volt), plus the 0.5 junction drop, or 1.5 volts. This voltage will be dropped across R_2.

The drop across R_1 must therefore be the total power-supply voltage (10 volts), minus the 1.5 volts at the junction, or 8.5 volts.

Using Ohm's law ($R = E/I$), $R_1 = 8.5/0.24$ mA $= 35,416$ ohms

$$R_2 = \frac{1.5}{0.24 \text{ mA}} = 6250 \text{ ohms}$$

With these values, the input signal to the amplifier could swing positive and negative as much as 1.5 volts (or slightly less) without canceling the base-emitter forward bias. Input signals greater than 1.5 volts would drive the amplifier out of Class A operation.

8-5. Final Circuit

A complete circuit showing the calculated resistance values and estimated voltage values is given in Fig. 8-2. The voltage values should be within 20 per cent of the estimated values.

The last step in design is to calculate the values for the two coupling capacitors C_1 and C_2, as well as the bypass capacitor C_3.

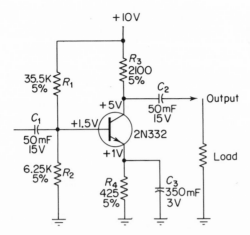

Fig. 8-2. Completed amplifier circuit using Current Design Method.

The values of these capacitors depend upon the *lowest* frequency with which the amplifier is to operate. Assuming that the lowest frequency is an arbitrary 10 Hz (which is adequate for all but special purpose amplifiers), the value of C_3 is calculated as follows:

Determine the value of the total resistance through which the capacitor must charge and discharge. Divide this value (in kilohms) into 150. The result should be the capacitance value in microfarads.

The value of emitter resistor R_4 is 425 ohms, or 0.425K. 150 divided by 0.425 = 353 microfarads (probably an electrolytic).

The total resistance value with which the coupling capacitors C_1 and C_2 are associated would have to be estimated. 1000 ohms is a good approximation for the input resistance of such an amplifier stage.

In the case of output coupling capacitor C_2, the charge and discharge would be through the input resistance of the following stage (estimated 1000 ohms), and the load resistance R_3 (approximately 2000 ohms), or a total of 3K ohms. 150 divided by 3K = 50 microfarads.

In the case of input coupling capacitor C_1, it is assumed that the preceding stage also has an output or load resistance of 2K, which is added to the 1K input resistance to provide 3K. Thus, the value of C_1 is the same as for C_2.

The capacitance values determined thus far will provide very good low-frequency response. If such response were not required, and the amplifier low limit were raised to say 100 Hz, the capacitance values could be cut in half without serious loss.

The working voltage values for the three capacitors can be estimated by using a minimum of 1.5 times the highest associated voltage. In the case of C_1 and C_2, the highest voltage is 10 volts, and a 15-volt (working) capacitor should be satisfactory. Normally, the greatest drop across C_3 is 1 volt, so a 1.5-volt capacitor will work satisfactorily. (However, from a practical standpoint, most electrolytics will have a 3-volt minimum working voltage).

CHAPTER 9

Controlled Rectifiers

The controlled rectifier is similar to the basic diode, with one specific exception. The controlled rectifier must be "triggered" or "turned on" by an external voltage or signal. The controlled rectifier has a high forward and reverse resistance (no current flow), without the trigger. When the trigger is applied, the forward resistance drops to zero (or very low), and a high forward current flows just as with a basic diode. The reverse current remains high, and no reverse current flows, so the controlled rectifier will rectify a-c in the normal manner.

So long as the forward voltage is applied, the forward current will continue to flow. The forward current will stop, and the controlled rectifier will "turn off," if the forward voltage is removed.

There are a great number of controlled rectifiers in use. Many of these are actually the same type (or a slightly different version), but manufactured under a different trade name or designation. This chapter discusses the basic types of controlled rectifiers, using the names and designations in most common use.

9-1. Silicon- or Semiconductor-Controlled Rectifier (SCR)

With some manufacturers, the letters SCR refer to *semiconductor-controlled rectifier*, and can mean any type of solid-state controlled rectifier. However, SCR usually refers to *silicon-controlled rectifiers*.

If four semiconductor materials, two *P*-type and two *N*-type, are arranged as shown in Fig. 9-1(a), the device can be considered as three diodes arranged alternately in series, as shown in Fig. 9-1(b). Such a device acts as a conventional diode rectifier in the reverse direction, and as an electronic switch and a rectifier in series in the forward direction. Conduction in the forward direction can be controlled or "gated" by operation of the switch.

Figure 9-2 shows the circuit symbol, block diagram, and basic physical construction of a typical SCR. Note that there are two symbols: one with the gate terminal connected to the cathode, and one to the anode. The cathode gate is the most common.

SCRs are most often used to control alternating current, but can be used to control a direct current. Either a-c or d-c can be used as the gate signal, providing the gate voltage is large enough to trigger the SCR into the "on" condition.

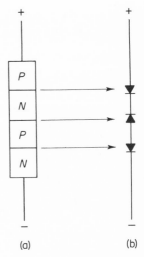

(a)   (b)

Fig. 9-1. Semiconductor material arrangement and equivalent diode relationship of an SCR.

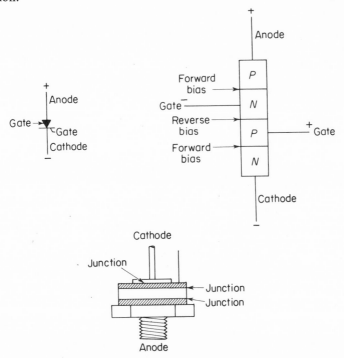

Fig. 9-2. Symbol, block diagram, and basic construction of a typical SCR.

An SCR is used to best advantage when both the load and trigger are alternating current. With a-c, control of the power applied to the load is determined by the relative phase of the trigger signal versus the load voltage. Because the trigger control is lost once the SCR is conducting, an a-c voltage at the load permits the trigger to regain control. Each alternation of a-c through the load causes conduction to be interrupted (when the a-c voltage drops to zero between cycles), regardless of the polarity of the trigger signal.

Figure 9-3 shows operation of an SCR with a-c voltages at the trigger circuit and load circuit. If the trigger voltage is in phase with the a-c power input signal as shown in Fig. 9-3(b), the SCR will conduct for successive positive alternations at the anode. When the trigger is positive-going at the same time as the load or anode voltage, load current starts to flow as soon as the load voltage reaches a value which will cause conduction. When the

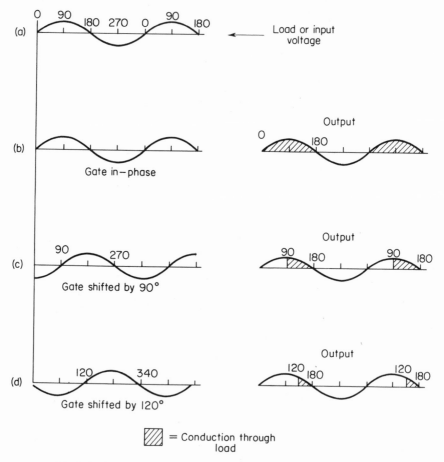

Fig. 9-3. Operation of an SCR with a-c at both load and gate or trigger.

trigger voltage is negative-going, the load voltage is also negative-going, and conduction stops. The SCR acts as a half-wave rectifier in the normal manner.

If there is a 90-degree phase difference between the trigger voltage and load voltage (say load voltage lags trigger voltage by 90 degrees) as shown in Fig. 9-3(c), the SCR will not start conducting until the trigger voltage swings positive, even though the load voltage is initially positive. When the load voltage drops to zero, conduction will stop, even though the trigger voltage is still positive.

If the phase shift is increased between trigger and load voltages as shown in Fig. 9-3(d), conduction time will be even shorter, and less power will be applied to the load circuit. By shifting the phase of the trigger voltage in relation to the load voltage, it is possible to vary the power output, even though the voltages are not changed in strength.

9-2. PNPN Controlled Switch

The *PNPN* controlled switch (sometimes referred to as an SCS, silicon-controlled switch, or semiconductor-controlled switch) is similar in operation to an SCR. However, the SCS is a *PNPN* device where *all four* semiconductor regions are made accessible by means of terminals.

Figure 9-4 shows the circuit symbols, block diagram, and basic physical construction of a typical SCS. Note that two circuit symbols are shown, and that both are in common use by manufactuerers. Often, SCS will be used as SCR, with the extra gate terminal not connected.

In some applications, SCS can be considered as a transistor and diode in series. Figure 9-5 shows this arrangement. If a negative load voltage were applied to terminal 4, with the positive at terminal 1, the SCS would not turn on, no matter what trigger signals were applied. However, with a positive at terminal 4 and a negative at terminal 1, the SCS would be turned on by either a positive voltage at terminal 2 or a negative voltage at terminal 3.

The SCS also has the ability to turn off by means of a gate signal. The arrangement is shown in Fig. 9-6. It should be noted that the gate turn-off method will apply only when the conducting current is below a certain value.

The SCS is often considered as two transistors (an *NPN* and a *PNP*) connected as shown in Fig. 9-7. Both transistors are connected in common emitter circuits with the collector output of the *NPN* feeding into the base input of the *PNP*, and vice versa. If a positive trigger voltage is applied to the *NPN* base, the emitter-base junction will be forward biased, and some current will flow in the emitter-collector circuit. Since the *NPN* collector feeds the *PNP* base, the *PNP* will also become forward biased, and the *PNP* collector output will feed into the *NPN* base, adding to the trigger voltage.

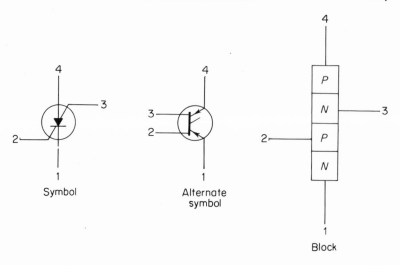

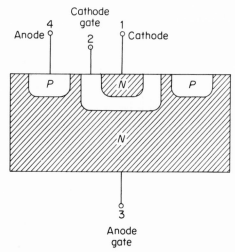

Fig. 9-4. Circuit symbols, block diagram, and basic physical construction of typical SCS (or PNPN controlled switch).

Load current then flows through the two transistors from the *NPN* emitter (or rectifier cathode) to the *PNP* emitter (or rectifier anode). Load current continues to flow even though the trigger is removed, since the current flow also keeps both transistors in the forward-biased condition.

The same condition can be produced by a negative trigger applied to the *PNP* base. This forward biases the *PNP* which, in turn, forward biases the *NPN*, until both transistors are conducting.

Normally, the currents will not stop until the load voltage is removed or, in the case of a-c, the voltage drops to zero between cycles. However, if

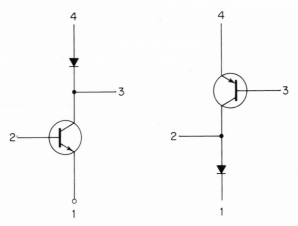

Fig. 9-5. Equivalent diode-transistor representation of an SCS.

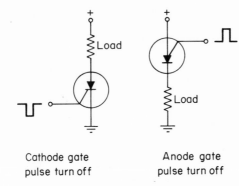

Cathode gate
pulse turn off

Anode gate
pulse turn off

Fig. 9-6. Gate turn-off arrangement for SCSs.

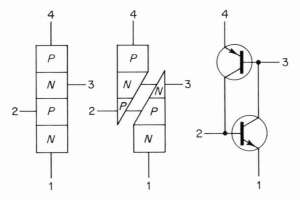

Fig. 9-7. Equivalent two-transistor representation of an SCS.

the load current is below a certain level (different for each type of SCS), the SCS can be turned off by a trigger voltage. For example, if a negative trigger voltage is applied to the *NPN* base, the emitter-base junction will lose some of its forward bias, and less current will flow in the *NPN* emitter-collector circuit. As a result, the *PNP* will also lose some of its forward bias, and the *PNP* collector output will drop to aid the turn-off trigger voltage at the *NPN* base.

The feedback process continues until the load current is completely stopped. It should be noted that the gain obtained during turn-off is much less than that during turn-on, since the turn-off current must buck the existing forward bias produced by the load current.

9-3. Triacs and Diacs

The Triac is a term coined by General Electric to identify the *tri*ode (three-terminal) *a-c* semiconductor switch. Like the SCR and SCS, the Triac is triggered by a gate signal. Unlike either the SCR or SCS, the Triac conducts in both directions, and is therefore most useful with a-c circuits. Since an SCR or SCS is essentially a rectifier, two SCRs or SCSs must be

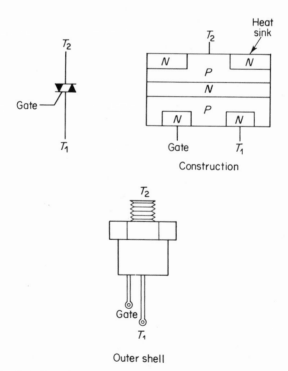

Fig. 9-8. Symbol, outer shell, and construction of a typical Triac.

connected back-to-back (in parallel or bridge) to control a-c. (Otherwise, the a-c will be rectified into d-c.) The use of two or more SCRs or SCSs requires elaborate control circuits (in many cases). Much of this can be eliminated by a Triac circuit when a-c is to be controlled.

Figure 9-8 shows the circuit symbol, outer shell, and basic physical construction of a Triac. Note that the symbol is essentially one SCR symbol combined with another complementary SCR symbol. Since the Triac is not a rectifier (when conducting, current flows in both directions), the terms "anode" and "cathode" do not apply. Instead, terminals are identified by number. Terminal T_1 is the reference point for measurement of voltages and currents at the gate terminal, and at terminal T_2. The area between terminals T_1 and T_2 is essentially a *PNPN* switch in parallel with an *NPNP* switch.

Like the SCS and SCR, the Triac can be made to conduct when a "breakdown" or "breakover" voltage is applied across terminals T_1 and T_2, and when a trigger voltage is applied. Current will continue to flow in one direction until that half-cycle of the a-c voltage (across T_1 and T_2) is complete. Current will then flow in the opposite direction for the next half-cycle. The Triac will not conduct on either half-cycle unless a gate-trigger voltage is present during that half-cycle (unless the breakdown voltage is exceeded).

The Triac can be triggered from many sources. One of the most common trigger sources is the Diac, shown in Fig. 9-9. Developed by General Electric, the Diac is a semiconductor device resembling a pair of diodes connected in complementary (parallel) form. The anode of one diode is connected to the cathode of the other diode, and vice versa.

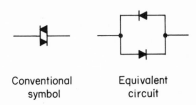

Conventional   Equivalent
symbol   circuit

Fig. 9-9. Symbol and equivalent circuit of Diac (*courtesy General Electric*).

Each diode passes current in one direction only, as in the case of a common diode. However, the diodes in a Diac will not conduct in the forward direction until a certain "breakover" voltage is reached. For example, if a Diac is designed for a breakover voltage of 3 volts, and the Diac is used in a circuit with less than 3 volts, the diodes appear as a high resistance (no current flow). Both diodes will conduct in their respective forward bias directions when the voltage is raised to any value over 3 volts.

9-4. Thyristor Parameters

The term "Thyristor" is applied by many manufacturers to SCS, SCR, *PNPN* switches, Triacs, and any number of similar semiconductor devices used for control. Likewise, each manufacturer has his own set of symbols, letters, and terms to identify the parameters of Thyristors. Many of the symbols and terms are duplicates used by different manufacturers. In a few cases, special terms and letter symbols are used by a manufacturer to identify the parameters of his own particular type of Thyristor. No attempt will be made to duplicate the data here.

However, the following paragraphs discuss the most important parameters. These parameters can then be compared with those found on the data sheets of a particular controlled rectifier.

Forward voltage is the voltage drop between the anode and cathode at any specified *forward anode current*, when the device is in the "on" condition. See Fig. 9-10.

Forward anode current is any value of positive current which flows through the device when it is in the "on" condition. See Fig. 9-10.

Forward "off" current is the anode current which flows when the device is in the "off" condition with a positive anode voltage applied. See Fig. 9-11.

Reverse anode voltage is any value of negative voltage which may be applied to the anode. The rated reverse anode voltage is less than the reverse

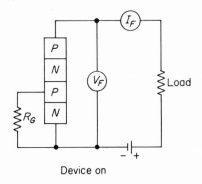

Device on

Fig. 9-10. Forward voltage and forward anode current test circuit.

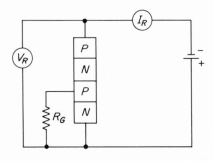

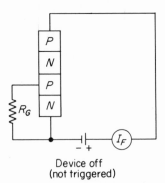

Device off
(not triggered)

Fig. 9-11. Forward "off" current test circuit.

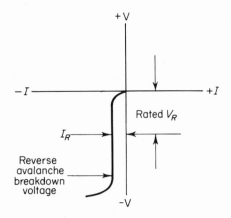

Fig. 9-12. Reverse anode voltage and reverse current test circuit.

avalanche bias voltage, and is the maximum peak inverse voltage of the device. See Fig. 9-12.

Reverse current is the negative current which flows through the device at any specified condition of *reverse anode voltage*, and temperature. Figure 9-12 shows the reverse current, at the rated reverse anode voltage.

Forward gate voltage is the voltage drop across the gate-to-cathode junction at any specified forward gate current. See Fig. 9-13.

Forward gate current is any value of positive current which flows into the gate, with a shunt resistance (R_G) between the gate and cathode. See Fig. 9-13.

Reverse gate voltage is any value of negative voltage which may be applied to the gate. Rated reverse gate voltage is less than the reverse breakdown voltage of the gate-to-cathode junction and is the maximum peak inverse voltage of the gate. See Fig. 9-14.

Forward breakover voltage is the value of positive anode voltage at which a controlled rectifier will switch into the "on" state, with a shunt resistance between the gate and cathode. Forward breakover voltage may also be considered the forward avalanche voltage. See Fig. 9-15.

Holding current is the minimum anode current required to sustain the device in the "on" condition. See Fig. 9-15.

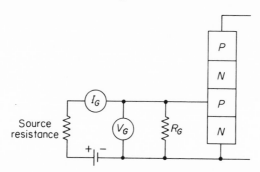

Fig. 9-13. Forward gate voltage and forward gate current circuit.

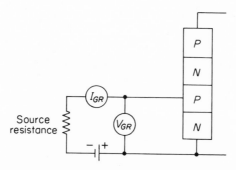

Fig. 9-14. Reverse gate voltage test circuit.

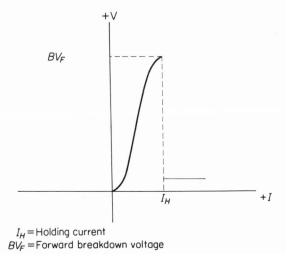

I_H = Holding current
BV_F = Forward breakdown voltage

Fig. 9-15. Typical curve showing forward breakdown voltage and holding current.

Junction temperature, in a *PNPN* device, is usually considered to be a composite temperature of all three junctions. Since all parameters of the device are temperature dependent, its operating junction temperature is very important.

Delay time is the interval between the start of the gate pulse and the instant at which the output has changed to 10 per cent of its maximum amplitude. In Fig. 9-16, this is the time, following initiation of the gate pulse, required for the anode voltage to drop to 90 per cent of its initial value. Delay time decreases with increased gate current, but at large gate currents reaches a limit of about 0.2 microsecond.

Rise time is the interval during which the output pulse changes from 10 per cent to 90 per cent of its maximum amplitude. In Fig. 9-16, this is the time required for the anode-to-cathode voltage to drop from 90 per cent to 10 per cent of its initial value.

Turn-on time is usually expressed as a combination of *delay time* plus *rise time*.

Turn-off time is the interval between the start of the turn-off and the instant at which the anode voltage may be reapplied, without turning on the unit. When a Thyristor has been triggered, and the anode voltage is positive, the unit will conduct. When the anode voltage swings negative, conduction stops. However, if the anode voltage were to swing positive immediately, the unit could conduct even though the trigger were not present. There must be some delay between the instant the conduction stops, and the instant the anode can be made positive. This delay is the turn-off time.

There are a number of factors that affect turn-off time. These include:

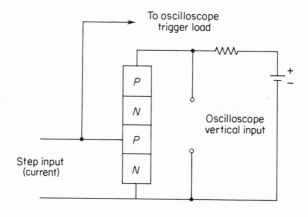

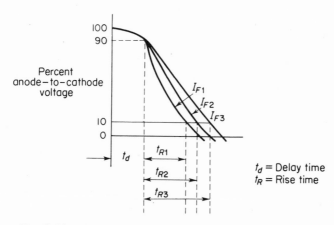

Fig. 9-16. Delay, rise, and turn-on time test circuit and graph.

Junction temperature. Turn-off time increases with increase in junction temperature.

Forward current and its rate of decay. Turn-off time increases as forward current and its rate of decay increase.

Reverse recovery current. If the device is subjected to a reverse bias (anode made negative) immediately after a condition of forward conduction (such as occurs when a-c is used, and the anode swings from positive to negative each half-cycle), a reverse (or recovery) current will flow from anode to cathode. This is essentially the same as recovery current of a conventional diode. Turn-off time will decrease as reverse (or recovery) current increases.

Rate of rise of reapplied forward voltage and its maximum amplitude. As the rate of rise and the amplitude of reapplied forward voltage increase, turn-off time increases.

Rate-of-rise (or dV/dT). When a rapidly rising voltage is applied to the anode of a Thyristor, the anode may start to conduct, even though there is no trigger, and the breakdown voltage is not reached. This condition is known as *rate effect*, or *dV/dT effect*, or (sometimes) *dI/dT effect*. The letters *dV/dT* signify a difference in voltage for a given difference in time. The letters *dI/dT* signify a difference in current for a given difference in time. Either way, the letter combinations indicate how much the voltage (or current) changes for a specific time interval. Usually, the terms are expressed in volts of change per microsecond, or amperes of current per microsecond.

Every Thyristor will have some *critical rate-of-rise*. That is, if the voltage (or current) rises faster than the critical rate-of-rise value, the unit will be turned on (with or without a trigger) even though the actual anode voltage does not exceed the rated breakdown voltage.

The critical rate-of-rise characteristic is especially important where a pulse-type signal, rather than a sine wave voltage, is applied across the anode.

9-5. Basic Controlled Rectifier Tests

As in the case of diodes and transistors, controlled rectifiers are subjected to many tests during manufacture. Few of these tests need be duplicated in the field.

One of the simplest and most comprehensive tests for a controlled rectifier is to operate the unit in a circuit that simulates actual circuit conditions (a-c and an appropriate load at the anode, a-c or pulse at the gate), and then measure the resulting conduction angle on a dual-trace oscilloscope.

With such an arrangement, the trigger and anode voltages, as well as the load current, can be adjusted to normal (or abnormal) dynamic operating conditions, and the results noted. For example, the trigger voltage can be adjusted over the supposed minimum and maximum trigger levels. Or the trigger can be removed, and the anode voltage raised to the actual breakover. The conduction angle method should test all important characteristics of a controlled rectifier, except for turn-on, turn-off, and rate-of-rise (which are discussed in later paragraphs of this chapter).

Conduction Angle Test

A dual-trace oscilloscope can be used to measure the conduction angle of a controlled rectifier. As shown in Fig. 9-17, one trace of the oscilloscope displays the anode current, while the other trace displays the trigger voltage. Both traces are voltage-calibrated. The anode load current is measured through a one-ohm, noninductive resistor. The voltage developed across this resistor is equal to the current, since $I = E/R$, and $R = 1$. For example, if a 3-volt indication is obtained on the oscilloscope trace, a 3-ampere current

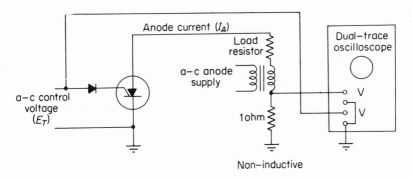

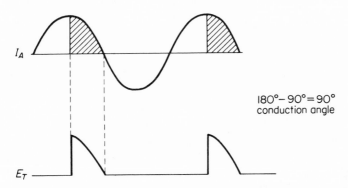

Fig. 9-17. Measurement of conduction angle with an oscilloscope.

is flowing in the anode circuit. The trigger voltage is read out directly on the other oscilloscope trace. Note that a diode has been placed in the trigger circuit to provide a pulsating d-c trigger. This can be removed if desired. Since the trigger is synchronized with anode current, the portion of the trigger cycle in which anode current flows is the conduction angle.

(1) Connect the equipment as shown in Fig. 9-17.

(2) Place the oscilloscope in operation as described in the applicable handbook. Switch on the internal recurrent sweep. Set the sweep-selector and sync-selector to internal.

(3) Apply power to the controlled rectifier. Adjust the trigger voltage, anode voltage, and anode current to the desired levels. Anode voltage can be measured by temporarily moving the oscilloscope probe (normally connected to measure gate voltage) to the anode.

(4) Adjust the oscilloscope sweep frequency and sync controls to produce two or three stationary cycles of each wave on the screen.

(5) On the basis of one conduction pulse equalling 180°, determine the angle of anode current flow, by reference to the trigger voltage trace. For example, in the display of Fig. 9-17, anode current starts to flow at 90° and stops at 180°, giving a conduction angle of 90°.

NOTE

If the unit under test is a Triac (or similar device) there will be a conduction display on both half-cycles.

(6) To find the minimum or maximum required trigger level, vary the trigger voltage from zero across the supposed operating range, and note the level of the trigger voltage when anode conduction starts.

(7) To find the breakdown voltage, remove the trigger voltage, and move the oscilloscope probe to the anode. Increase the anode voltage until conduction starts, and note the anode voltage level.

Rate-of-Rise Tests

Various manufacturers have developed a number of circuits for rate-of-rise tests. All of these circuits are based on a method known as the *Exponential Waveform Method*. The following is a description of the basic circuit and techniques for the method.

The basic test circuit is shown in Fig. 9-18. In operation, a large capacitor C_1 is charged to the *full voltage rating* of the unit under test. Capacitor C_1 is then discharged through a variable time-constant network (R_2 and C_2). This is repeated with smaller time constants (higher dV/dT) until the unit under test is turned on by the fast dV/dT.

The critical rate, which causes firing, is defined as:

$$dV/dT = \frac{0.632 \times \text{anode voltage}}{R_2 \times C_2}$$

This equation describes the average slope of the essentially linear rise portion of the applied voltage. See Fig. 9-19.

In practical test circuits, there are two major conditions which determine the value of the circuit components:

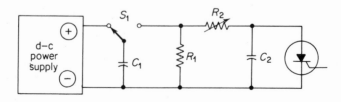

Fig. 9-18. Basic test circuit for rate-of-rise (dV/dT).

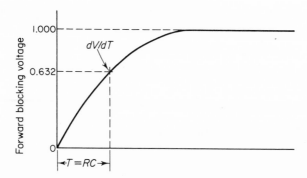

Fig. 9-19. Exponential applied forward voltage and definition of dV/dT.

First, capacitor C_1 should be large enough to serve as a constant voltage source during the discharging of C_1 and the charging of C_2.

Second, capacitor C_2 should be much larger than the intrinsic cathode-to-anode capacitance of the unit under test, plus any stray device and device-test wiring capacitances. Controlled rectifier-junction capacitance is a diminishing function of the applied voltage, and has its highest value for zero anode-to-cathode voltage.

Typically, for 70-ampere devices, the junction capacitance has been found to be in the order of 800 pF for zero applied voltage.

Generally, 0.5 microfarad for C_1, and 0.01 microfarad for C_2, are practical values.

It should be noted that the stray inductance and capacitance of the test circuit should be minimized. This is especially true for measurement of high dV/dT values.

Turn-On and Turn-Off (Recovery) Time Tests

Figure 9-20 shows a circuit capable of measuring both turn-on and turn-off (recovery) time. The circuit inductances must be kept to a minimum by using short connections, thick wires, and closely spaced return loops or wiring on a grounded chassis. External pulse sources must be provided for the circuit. These pulses are applied to transformers T_1 and T_2, and serve to turn the unit under test on and off. The pulses can come from any source, but should be of the amplitude, duration, and repetition rate that correspond to the normal operating conditions of the unit under test.

When a suitable gate pulse is applied to transformer T_1, the unit under test is turned on. Load current can be set by resistor R_L. A predetermined time later, the turn-off controlled rectifier is turned on by a pulse applied to T_2. This places capacitor C_2 across the unit under test, applying a reverse bias, and turning the unit under test off.

Any oscilloscope capable of a 10-microsecond sweep can be used for viewing both the turn-on and turn-off action. The oscilloscope is connected

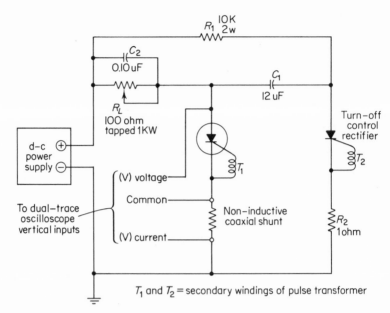

Fig. 9-20. Turn-on and turn-off (recovery) test circuit
(courtesy International Rectifier).

with the vertical input across the unit under test. Turn-on time is displayed when the oscilloscope is triggered with the gate pulse applied to the unit under test. Turn-off time is displayed when the oscilloscope is triggered with the gate pulse applied to the turn-off controlled rectifier.

The actual spacing between the turn-on and turn-off pulses is usually not critical. However, a greater spacing will cause increased conduction, and will heat the junction of the unit under test. Since operation of controlled rectifiers is temperature dependent, the rise in junction temperature must be taken into account for accurate test results. (Both turn-on and turn-off times increase with an increase in junction temperature.)

Figure 9-21 shows turn-on action. Turn-on time is equal to delay time (t_d), plus rise time (t_r). Following the beginning of the gate pulse, there is a short delay before appreciable load current flows. Delay time is the time from the leading edge of the gate-current pulse (beginning of oscilloscope sweep) to the point of 10 per cent load-current flow. (Delay time can be decreased by overdriving the gate.)

Rise time (t_r) is the time the load current increases from 10 per cent to 90 per cent of its value. Rise time depends upon load inductance, load-current amplitude, junction temperature, and, to a lesser degree, upon anode voltage. The higher the inductance and load current, the longer the rise time. An increase in anode voltage tends to decrease the rise time. Capacitor

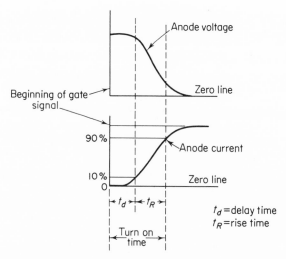

Fig. 9-21. Definition of turn-on time (delay time plus rise time) *(courtesy International Rectifier).*

C_2 (Fig. 9-20) tends to counter the load inductance, thereby lessening the rise time.

By triggering the oscilloscope with the gate pulse applied to the unit under test, the sweep starts at the gate pulse leading edge. Thus, the oscilloscope presentation shows the anode voltage from this point on. In noninductive circuits, when the anode voltage decreases to 90 per cent of initial value, this time is equal to 10 per cent of the load current, and is therefore equal to the delay time.

With the oscilloscope set at 10-microsecond sweep, each square represents 1 microsecond. The delay time is read directly by counting the number of divisions on the oscilloscope screen. If the circuit is *noninductive*, the decrease from 90 per cent to 10 per cent of the anode voltage is *approximately* equal to the load current increasing from 10 per cent to 90 per cent. The time this takes is equal to the rise time. The total time from zero time (start of oscilloscope sweep) to this 10 per cent of the anode voltage is equal to the turn-on time. Therefore, turn-on time is determined by counting the number of divisions from the start of the oscilloscope sweep to the 90 per cent anode-load current.

Figure 9-22 shows the reverse current and reverse recovery (turn-off) action of the unit under test. Turn-off time is the time necessary for the unit under test to turn off, *and recover its forward blocking ability.* The reverse recovery time (t_h) is the length of the interval between the time the forward current falls to zero when going reverse, and the time it returns back to zero from the reverse direction.

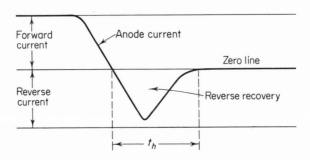

t_h=reverse recovery time

Fig. 9-22. Definition of reverse recovery time
(*courtesy International Rectifier*).

In Fig. 9-20, the time available for turn-off action is determined by the value of capacitor C_1 and resistor R_2. Decreasing the value of C_1 decreases the time the unit under test is reverse biased. The resistor R_2 limits the magnitude of the reverse current. The shape of the reverse voltage and current pulses are given by a capacitor-resistor discharge. At the end of the reverse pulse, forward voltage is reapplied. Having turned off, the unit under test blocks forward voltage, and no current can flow.

Figure 9-22 shows the reverse current pulse. With the oscilloscope set at 20-microsecond sweep, the value of reverse recovery time (turn-off time) can be measured by counting off the divisions from the zero point on the leading edge of the reverse current pulse, to the zero point on the trailing edge.

CHAPTER 10

Photocell Data

This chapter discusses the practical aspects of photocells. In most cases, the laboratory technician is interested in the output of photocells in relation to the external circuit. That is, how much voltage (current, or power) will a photocell of a given size and type produce for a given load and amount of light? Or, what is the optimum load resistance for a given photocell operating in a given amount of light? Therefore, the bulk of this chapter will be devoted to such factors. However, the data of this chapter will also familiarize the reader with the basic operation of photocells.

10-1. Basic Photocell Theory

There are two basic types of photocells, *photovoltaic* and *photoconductive*.

The photovoltaic cells produce an output voltage and current in the presence of light. For this reason, photovoltaic cells are often termed *solar batteries* or solar cells.

The photoconductive cells are often termed *light sensitive resistors* since they function as a resistor, and do not generate an output. Instead, photoconductive cells act as a resistance which varies in the presence of light, thus changing the amount of current being conducted through the circuit.

10-2. Photovoltaic Cell Theory

A *silicon photovoltaic cell* is essentially a *PN* junction. Figure 10-1 shows the construction of a typical silicon photovoltaic cell, as well as the electrical symbol.

In manufacturing such junctions, wafers of *N*-type silicon are ground, and then subjected to a chemical such as boron. The boron diffuses into the surface and creates a thin, *P*-type silicon layer. The *P*-type layer is then removed from all but one side of the wafer, and contacts are added. The *P*-type layer is made thin so that light may pass through it to the junction.

When light rays strike the silicon atoms near the junction, electrons are agitated and liberated from the atoms. These electrons move about, leaving behind positively charged holes. In the process, a *contact potential* or *space charge* appears across the junction. (This condition is discussed further in Chapter 11.) As a result of the electron movement, a current flows, with the *N*-layer of silicon acting as the negative terminal of the "battery."

The brighter the light, the more electrons that are set free, and the more current that will flow in the external circuit connected to the cell terminals. Output current is directly proportional to the total light that falls on the surface of the cell.

Usually, photovoltaic cells are rated as to a certain output voltage for a given amount of light (specified in *footcandles*).

Figure 10-2 shows the equivalent circuit of a photovoltaic cell. R_{shunt} is the internal shunt leakage path, C is the inherent capacitance, R_{series} is the series resistance due to the transparent conducting layer, and the lead connections, and R_{load} is the external load. R_{series} is a fixed value. R_{shunt} is

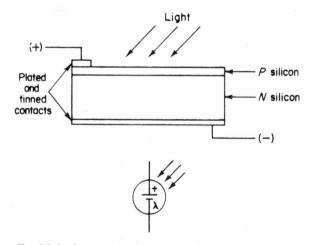

Fig. 10-1. Construction of a typical silicon photovoltaic cell.

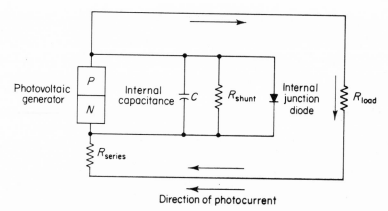

Fig. 10-2. Equivalent circuit of photovoltaic cell.

variable, its value depending on the area of the photovoltaic cell, and on its illumination. The larger the area or the higher the illumination, the lower is the value of R_{shunt}. Under these conditions, a lower value of R_{load} would be required for maximum power transfer. The value of C increases with increasing area, and decreases the response to rapidly flickering light.

Selenium photovoltaic cells, like silicon, produce electrical current when illuminated. Although selenium is not as efficient as silicon, selenium solar cells have characteristics more nearly like the human eye. That is, selenium is highly efficient in the range of human vision, but not beyond, as is silicon. Therefore, selenium is used in applications where the light sources are similar in level to those that would normally be seen by the human eye.

Figure 10-3 shows the construction of a typical selenium photovoltaic cell, as well as the electrical symbol. Note that the symbols for silicon and selenium photovoltaic cells are the same.

Selenium photovoltaic cells are made by depositing a thin film or layer

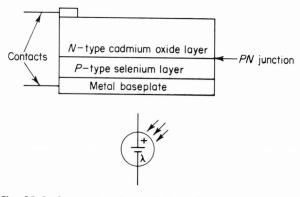

Fig. 10-3. Construction of a typical selenium photovoltaic cell.

of selenium onto a metal base plate. The selenium is then crystalized by heat. This is followed by depositing cadmium oxide on the selenium to form the junction. Note that the selenium layer forms the *P*-region, while the cadmium oxide forms the *N*-region.

10-3. Photoconductive Cell Theory

The photoconductive cell is essentially a resistor rather than a *PN* junction. Figure 10-4 shows the construction of a typical photoconductive cell, as well as the electrical symbol. Note that the symbol for a photoconductive cell does not resemble the symbol for the photovoltaic cell. The photoconductive cell is basically a resistance that varies with changes in light, while the photovoltaic cell is basically a battery that varies with light changes.

A typical photoconductive cell consists of a glass envelope containing a plate of photoconductive material (usually cadmium sulphide). Metal film electrodes form two interlocking combs which make up one side of the photoconductive coating. These comb-like electrodes increase the light sensitivity of the cell, and enlarge the area of sensitivity.

The basic circuit for a photoconductive cell is shown in Fig. 10-5. An external voltage source is required since photoconductive cells do not produce a voltage of their own. When the external voltage is applied, the photoconductive cell appears as a resistance. When light rays strike the photoconductive surface, electrons are agitated and liberated from their atoms. These electrons move about, leaving behind positively charged holes. The

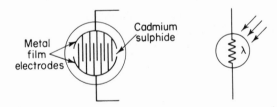

Fig. 10-4. Construction of typical photoconductive cell.

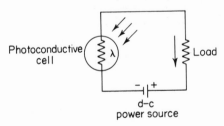

Fig. 10-5. Basic circuit for photoconductive cell.

increased flow of electrons causes a corresponding increase in current (or a decrease in resistance), even though the external voltage does not change.

When a photoconductive cell is connected as shown in Fig. 10-5, a voltage will be developed across the load resistor. This voltage will be directly proportional to current, and to the available light striking the cell. If light increases, the cell resistance will decrease, and more current will flow. This will cause a greater voltage drop across the load resistor.

The *photoconductive transistor* is another form of photoconductive device. Figure 10-6 shows the construction of a typical germanium photoconductive transistor, as well as the electrical symbol. Note that the symbol for a photoconductive transistor is similar to that of a conventional transistor. Many times, a standard transistor symbol (together with arrowheads) will be used to indicate a photoconductive transistor.

The photoconductive transistor (or *phototransistor* as it is sometimes called) operates on the principle that germanium is a photoconductive material. This permits the use of a very small photocell which operates on the *point contact* transistor principle, rather than on the usual *PNP* or *NPN* junction. Instead of an emitter acting as an input circuit (as is the case for a common-base transistor) the emitter input is a beam of light which causes current to flow.

As shown in Fig. 10-6, a concave-shaped germanium element is mounted in a metal shell which forms the base connection. The current flow from the negative side of the battery goes through the load resistor on to the collector, and then to the base. Like the cadmium sulphide cell, the phototransistor does not generate its own electrical current, but must have an external voltage applied. The phototransistor has only an extremely small area of light-sensitive surface (compared with the typical cadmium sulphide cell). The light sensitive area centers around the section of the base which is opposite the collector contact point.

Both the photocell and phototransistor are used as *light actuated switches*

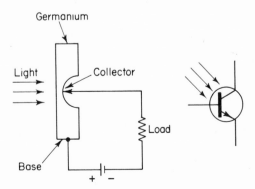

Fig. 10-6. Construction of photoconductive transistor.

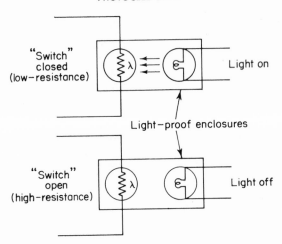

Fig. 10-7. Typical light-actuated switch arrangement.

(LAS), also known as *light sensitive* or *light dependent switches*. When the photocell is exposed to a very strong light, the internal resistance drops to zero (or near zero). This has the same effect as closing the contacts of a switch. When all light is removed from the cell or transistor, the internal resistance increases to several million ohms. This has the same effect as opening the switch contacts.

Semiconductor light-actuated switches have many applications. An example is shown in Fig. 10-7 where a photoconductive cell and a lamp are sealed in a light-proof enclosure. When the lamp is extinguished, the cell resistance is very high, and the "switch" is "open." When current is applied to the lamp, the light strikes the cell, and causes the cell resistance to drop to zero. This "closes" the "switch." Such a switch is completely noise-free, an important factor in many computer applications.

10-4. Photovoltaic Output Characteristics

The output characteristics (voltage, current power) for silicon and selenium photovoltaic cells are usually given on the data sheets. Photocell data sheets also often specify optimum load resistances. Sometimes output data are given in the form of tables, while other data sheets use graphs.

Figures 10-8 and 10-9 are typical examples of the tabular method, showing output characteristics for selenium and silicon cells, respectively.

Figures 10-10 and 10-11 are typical examples of the graph method. Figure 10-10 shows the relationship between output current, illumination, and load resistance for a selenium photovoltaic cell with a 1-square-inch active area. Figure 10-11 shows the relationship between output current, illumination, and active cell area for given load resistance values.

Hermetically sealed photocells

Hermetically sealed photocells are protected against humidity, corrosive atmospheres, and salt spray. These cells are especially suitable for outdoor applications, where protection from corrosion is required, and may be immersed in non-corrosive liquids.

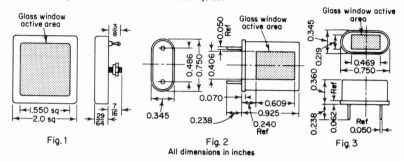

Fig. 1   Fig. 2   Fig. 3

All dimensions in inches

Hermetic sealed cell type	Fig	Photosensitive area inches²	Typical output current			Typical output voltage	
			100 fc 100 ohms microamps	100 fc 1000 ohms microamps	1 fc 10,000 ohms microamps	100 fc RL=1 megohm volts	1 fc RL=100,000 ohms millivolts
DP-5	1	2.25	600	250	3.5	0.30	75
DP-3	2	0.21	66	60	0.6	0.26	34
DP-2	3	0.088	24	20	0.16	0.22	9

Mounted photocells in plastic housings

The cells are mounted in a black phenolic or butyrate housing with glass or plastic window. The mounting studs protruding at the rear of the housing also serve as the electrical output terminals.

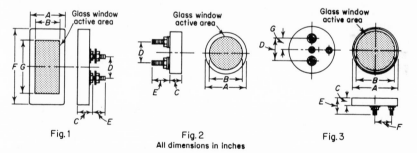

Fig. 1   Fig. 2   Fig. 3

All dimensions in inches

Mounted cell type	Fig	Dimensions							Typical output current			Typical output voltage	
		A	B	C	D	E	F	G	100 fc 100 ohms microamps	100 fc 1000 ohms microamps	1 fc 10,000 ohms microamps	100 fc 1 megohm volts	1 fc 100,000 ohms millivolts
B10-M	1	1.1	0.7	0.4	0.6	0.4	2.2	2.2	320	210	2.0	0.30	65
A5-M	2	1.2	1.1	0.4	0.6	0.4			220	160	1.5	0.30	60
A7-M	2	1.7	1.4	0.4	0.6	0.4			350	200	2.9	0.31	60
A10-M	2	1.9	1.7	0.4	0.6	0.4			550	240	3.4	0.30	70
A15-M	3	2.1	1.9	0.4	1.0	0.4	0.8	0.5	700	250	3.5	0.31	70

Fig. 10-8. Typical selenium cell output characteristics
(courtesy International Rectifier).

163

MODULES

These silicon photovoltaic cells are designed and manufactured to rigid military specifications, and utilize the most advanced techniques in semi-conductor technology. Due to their extremely long life, high conversion efficiency (to 12%), wide operating temperature range, simplicity of operation and high power to weight ratio, silicon solar cells offer an extremely efficient method for the direct conversion of solar energy.

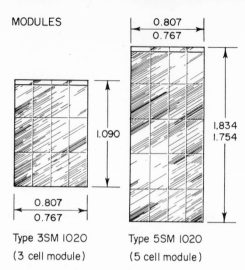

Type 3SM 1020
(3 cell module)

Type 5SM 1020
(5 cell module)

STANDARD MODULE CELL TYPES
Typical characteristics for optimum power transfer at 26°C cell temperature and 100 mW/cm^2 solar irradiation

Module types (1)	Number of cells in series	Minimum conversion efficiency (%)	Minimum output power (mW)	Approximate output voltage Volts (2)	Approximate output current (mA)
2SM1020E4	2	4	14.4	0.70	20
2SM1020E5	2	5	18.0	0.72	25
2SM1020E6	2	6	21.6	0.74	29
2SM1020E7	2	7	25.2	0.76	33
2SM1020E8	2	8	28.8	0.78	37
3SM1020E4	3	4	21.6	1.05	20
3SM1020E5	3	5	27.0	1.08	25
3SM1020E6	3	6	32.4	1.11	29
3SM1020E7	3	7	37.8	1.14	33
3SM1020E8	3	8	43.2	1.17	37
5SM1020E4	5	4	36.0	1.75	20
5SM1020E5	5	5	45.0	1.80	25
5SM1020E6	5	6	54.0	1.85	29
5SM1020E7	5	7	63.0	1.90	33
5SM1020E8	5	8	72.0	1.95	37

Notes:

1. Modules of higher conversion efficiency are also available on request.

2. These factors vary slightly even with modules of the same efficiency.

3. For elevated cell temperature, use the correction factor from Fig.1 for calculation of optimum power output and voltage at optimum power output. The current at optimum power output does not vary appreciably with temperature (from 10 to 130°C).

4. For other illumination intensities than 100 mW/cm^2, use the appropriate correction factor from Fig. 4 to find the optimum power output, voltage at optimum power output and current at optimum power output.

5. All silicon solar cells and modules are available with pigtail lead terminals. Simply add the suffix "PL" to the standard type number when pigtail leads are desired.

6. For additional design data, contact the factory. Complete design recommendations will be supplied on request.

Fig. 10-9. Typical silicon cell output characteristics
(*courtesy International Rectifier*).

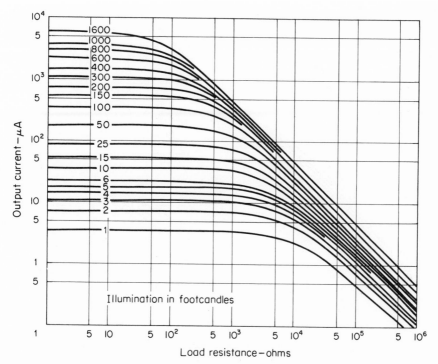

Fig. 10-10. Typical curves showing output characteristics of selenium cells *(courtesy International Rectifier).*

While it is essential that the information on photocell data sheets be followed, it is sometimes convenient to have some rules of thumb for matching load resistances to power requirements, and vice versa.

The following procedures (developed by International Rectifier Corporation) will provide basic rules for selecting photovoltaic power supplies.

Selecting Correct Load Resistance

If the photovoltaic cell is to operate in full sunlight (approximately 10,000 footcandles), divide 500 by the full current output (short circuit output) in milliamperes of the cell.

For example, an International Rectifier Corporation B2M selenium photovoltaic cell delivers a maximum (short circuit) output of 2 to 5 mA.

500 divided by 2 = 250, 500 divided by 5 = 100

Therefore, the optimum load resistance would be from 100 to 250 ohms.

A similar rule can be applied for other illuminations. The following Table 10-1 provides the number to use at illuminations from full sunlight (arbitrarily set at 10,000 footcandles) down to 1 footcandle.

Output vs. illumination
at fixed load resistance

These curves may aid in the selection of the proper cell area with the desired sensitivity for applications with an input resistance of the values shown.

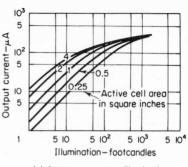

(c) Output current vs. illumination for 1500 ohms load resistance.

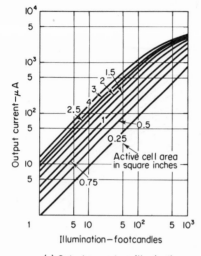

(a) Output current vs. illumination for 100 ohm load resistance

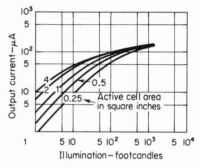

(d) Output current vs. illumination for 3000 ohms load resistance

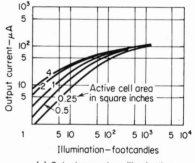

(e) Output current vs. illumination for 4000 ohms load resistance.

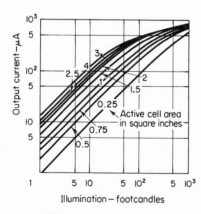

(b) Output current vs. illumination for 500 ohms load resistance.

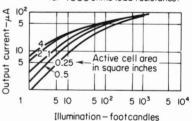

(f) Output current vs. illumination for 5000 ohms load resistance.

Fig. 10-11. Typical curves showing load resistance factors of photocells (*courtesy International Rectifier*).

TABLE 10-1

Factors for Determining Load Resistance of Photovoltaic Cells

Illumination (Footcandles)	Divide This Number by the Short Circuit Current in Milliamperes
10,000	500
1,000	400
100	300
10	200
1	100

When the short circuit output of a particular cell is not given on the data sheets, find the short circuit output by actual test, using a current meter, and a light source equivalent to that where the cell is to be used. Use a current meter (micrommeter or milliammeter) of the highest possible sensitivity, since any meter movement will have some resistance across the cell output.

For example, assume that a particular cell is to be operated at 100 footcandles, and there is no data-sheet information available for the particular cell. Place the cell in the 100-footcandle light, and measure the short circuit current. Assume that the short circuit current is 0.10 milliampere.

$$300 \text{ divided by } 0.10 = 3000 \text{ ohms}$$

Nominal Power Output

When a cell is operated in direct sunlight at sea level, the available solar energy is said to be 100 milliwatts per square centimeter, or 645 milliwatts per square inch. However, no cell is 100 per cent efficient, so the 100 mW/cm² must be multiplied by the efficiency rating of the particular cell.

Nominal Voltage Output

When a cell is operated in direct sunlight at sea level, the typical voltage per cell is 0.25 volt for selenium, and 0.4 volt for silicon.

Typical Cell Areas

In almost all applications for silicon, the area of the individual cell will be 0.9 by 2.0 centimeters, or 1.8 by 2.0 centimeters. For selenium, standard individual cell sizes are available up to about 6 by 6 inches.

Typical Optimum Load Resistances

The typical optimum load resistance, in direct noon sunlight at sea level, for 1 square inch of selenium photovoltaic cell is 10 ohms, and for a 1 by 2 centimeter silicon cell is 9 to 13 ohms.

Conversion of Square Inches and Square Centimeters

Area sizes of photocells are almost always given in square inches or square centimeters.

If square inches are given, multiply by 6.452 to find square centimeters.

If square centimeters are given, multiply by 0.155 to find square inches.

10-5. Photoconductive Output Characteristics

Although photoconductive cells do not have true power output characteristics (since such cells do not generate power), there are still operating characteristics which must be analyzed.

Usually, photoconductive cells are rated as to peak or operating voltage, power dissipation, and resistance at a given illumination.

For example, an RCA 7163 1-inch diameter cadmium-sulphide photoconductive cell is rated at a maximum (d-c or peak a-c) of 600 volts, a continuous service power dissipation of 750 milliwatts, and a resistance of 10K at 2 footcandles.

The major concern is that the power dissipation rating not be exceeded when the cell is exposed to maximum light. For example, assume that the resistance drops to 100 ohms in maximum light. The maximum voltage would be:

$$(E = \sqrt{P \times R})$$
$$E = \sqrt{0.750 \times 100} = 8.66 \text{ volts}$$

Obviously, the peak or maximum voltage rating can not be exceeded in any amount of light.

CHAPTER 11

Voltage-Variable
Junction-Diode Capacitors

An ordinary *PN* junction diode has all of the elements to form a capacitor. The *P*- and *N*-regions act as conductors and form the capacitor "plates." The junction between the two regions acts as a dielectric (unless the junction is forward biased). If a *PN* junction diode is reverse biased, the junction is reinforced to further prevent any current flow. This makes the diode act more like a capacitor where no current can flow between the plates (*P*- and *N*-regions), except for possible leakage current.

The fact that a diode will act like a capacitor is usually an undesired characteristic in conventional diode operation. However, the characteristic can be put to good use when it is desired to have a capacitor that will vary in value with changes in voltage.

Several manufacturers produce diode capacitors under such trade names as: Epicap, Capsil, Varactor, Voltacap, Paramp Diode, Semicap, Varicap. Likewise, there are a number of symbols used for diode capacitors as shown in Fig. 11-1. Usually, such capacitors are designated as *voltage-variable capacitors* or VVC.

Although any semiconductor material could be used for diode capacitors, silicon is used most often because of its high forward-to-reverse resistance ratio, and good temperature characteristics.

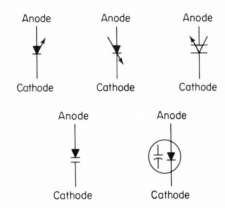

Fig. 11-1. Typical symbols for voltage-variable diode capacitors.

11-1. Operating Theory of Voltage-Variable Capacitors

When a junction is formed between N-type and P-type material, there is a transfer of charges across the junction. Holes from the P-region cross the junction to neutralize the "excess" electrons near the junction in the N-region. Likewise, electrons from the N-region cross the junction to neutralize positive carriers near the junction in the P-region. All free charged particles are removed from the immediate vicinity of the junction area due to the transfer. In the process, a *contact potential* or *space charge* (about 0.5 volt for silicon) appears across the junction. This condition is shown in Fig. 11-2.

A *PN* junction acts similarly to a slightly charged capacitor, with the

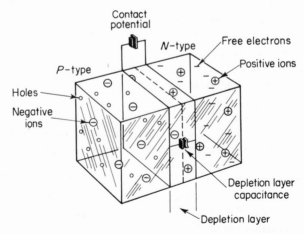

Fig. 11-2. Development of a space charge in a *PN* junction (*courtesy Motorola*).

depletion layer (or *depletion region*) representing the dielectric, and the semiconductor material adjacent to the depletion layer representing the two conductive plates.

If an external voltage is applied across the *PN* junction so as to reinforce the contact potential (that is, if a reverse bias is applied to the junction), the depletion layer is increased, and the capacitance is decreased. This condition is shown in Fig. 11-3.

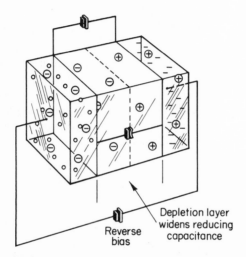

Depletion layer
widens reducing
capacitance

Reverse
bias

Fig. 11-3. *PN* junction space charge with reverse voltage
(*courtesy Motorola*).

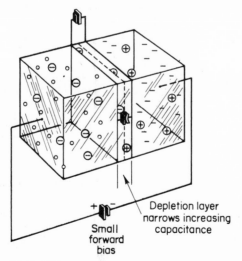

Depletion layer
narrows increasing
capacitance

Small
forward
bias

Fig. 11-4. *PN* junction space charge with forward voltage
(*courtesy Motorola*).

If a forward-bias voltage (or less reverse bias) is applied, the depletion layer decreases, and the capacitance is increased. This condition is shown in Fig. 11-4.

If a large forward bias is applied (sufficient to overcome the contact potential), forward conduction will occur, and the capacitance effect is destroyed. The *PN* junction then acts as a conventional forward-biased junction.

11-2. Equivalent Circuit of a Voltage-Variable Capacitor

In its simplest form, a voltage-variable capacitor can be represented as a capacitance in series with a resistance, as is shown in Fig. 11-5. However, as in the case of any *PN* junction, a voltage-variable capacitor must be protected from the effects of atmosphere, and is therefore housed in a package. There are parasitic reactances associated with the package and the internal connections to the junctions.

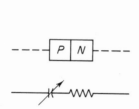

Fig. 11-5. Basic equivalent circuit of voltage-variable capacitor.

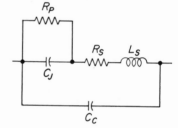

Fig. 11-6. Complete equivalent circuit of a voltage-variable capacitor (PN junction with reverse voltage).

The complete equivalent circuit of a packaged *PN* junction operated in the reverse voltage region (a voltage-variable junction-capacitor) is shown in Fig. 11-6. The voltage-variable capacitance is C_j; R_s is the series resistance; R_p is the junction-shunt resistance which generally can be neglected; L_s is the lead inductance, and C_c is the case or housing capacitance.

11-3. Characteristics of Voltage-Variable Capacitors

The following Table 11-1 indicates the characteristics needed to describe a voltage-variable capacitor for various applications.

Q, or *quality factor*, is the most important characteristic or parameter of voltage-variable capacitors.

TABLE 11-1

Typical Voltage-Variable Capacitor Characteristics

Motorola Number	Capacitance at −4 VDC (pf)	Q at 4 VDC 50MHz Min.	Standard Capacitance Tolerance at 4 VDC	Minimum Reverse Breakdown Voltage (10 microamperes)	Leakage Current at 48 Volts (microamperes)	Typical Capacity Change 0°C to 100°C Per Cent
MV1864B	6.8	600	±10%	60	0.5	2
MV1870	15	250	±10%	60	0.5	2
MV1872	22	200	±10%	60	0.5	2
MV1874	27	200	±10%	60	0.5	2
MV1876	33	200	±10%	60	0.5	2
MV1878	47	175	±10%	60	0.5	2

When only the series resistance R_s is concerned (Fig. 11-6), the equation is:

$$Q_s = \frac{1}{6.28f\,C\,R_s} \tag{11-1}$$

where:

Q_s is the quality factor (with series resistance only)
f is the frequency in Hz
C is the capacitance in farads
R_s is the series resistance in ohms

When the resistance is in parallel with the capacitance,

$$Q_p = 6.28f\,C\,R_p \tag{11-2}$$

where:

R_p is the parallel resistance in ohms

If both series and parallel resistances are present, the following is a simple equation for Q,

$$\frac{1}{Q} = \frac{1}{Q_s} + \frac{1}{Q_p} \tag{11-3}$$

where Q_s and Q_p are calculated by equations 11-1 and 11-2, respectively.

Tuning range is another important characteristic. With voltage-variable capacitors, a range of voltage swing must be determined that will allow maintenance of acceptable capacitor characteristics. The usual allowable voltage swing is from a few volts negative to the maximum reverse working voltage. *Capacity swing* is calculated from voltage swing by the inverse half-power relation between voltage and capacity. For large tuning ratios, the maximum-to-minimum voltage ratio should, obviously, be large.

The capacitance value of a voltage-variable capacitor can be calculated by:

$$C = \frac{C_o}{(1 + V/\Phi)^Y} \tag{11-4}$$

where:

  C is the capacitance at voltage V
  C_o is the capacitance at zero bias
  V is the external voltage across the junction (reverse bias)
  Φ is the contact potential (which is considered at 0.5 for silicon voltage-variable capacitors)
  Y is the power law of the junction (which is considered at 0.5 for silicon voltage-variable capacitors)

Temperature stability of the voltage-variable capacitor is also an important characteristic for most applications. This is usually expressed as a per cent of capacity change for a given change in temperature.

11-4. Determining Tuning Range of a Resonant Circuit Using Voltage-Variable Capacitors

When using a voltage-variable capacitor in any resonant circuit, the major concern is the tuning range of the circuit. All other circuit factors being equal, the tuning range of a resonant circuit is dependent upon the capacity range of the voltage-variable capacitor. A graph has been developed by Motorola in which tuning range can be predicted using the voltage-variable capacitance, and the external-circuit parameters.

Most voltage-variable capacitor (or VVC) resonant circuits are in the

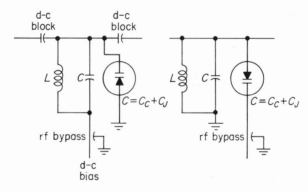

Fig. 11-7. Typical parallel circuits for VVC control *(courtesy Motorola)*.

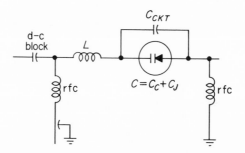

Fig. 11-8. Typical series circuit for VVC control (*courtesy Motorola*).

form of Fig. 11-7 for the parallel circuit, or Fig. 11-8 for the series circuit. The effective circuit inductance is given by L, although in some cases for biasing purposes there are additional *RF* chokes which, if properly chosen, have negligible effect on the resonant frequency. Circuit capacity shunting the VVC is given by C_{ckt}. The VVC capacity is given by $C_c + C_j$, the sum of case and junction capacitances.

NOTE

It is assumed that the resonant frequencies are well below the VVC self-resonant frequency so that any inductance can be ignored. It is not recommended that the VVC be operated in circuit near the self-resonant frequency.

Finding Capacity of a VVC When the Capacity at One Voltage Is Known

One of the steps in finding the tuning range of a resonant circuit (that uses a VVC) is to find the VVC capacity at various voltages. Most VVC specification sheets list the capacity at one voltage only. The capacity at other voltages can be calculated using the ratio of the two voltages. The basic equation is:

$$\text{Voltage ratio} = \left(\frac{1 + \text{known voltage}/0.5}{1 + \text{unknown voltage}/0.5} \right)^{1/2} \tag{11-5}$$

where:

> known voltage is the voltage where the capacity is known
> unknown voltage is the voltage where the capacity is not known

The equation of (11-5) also assumes a contact potential of 0.5, and a 0.5 power law of the junction (typical for silicon).

As an example: Assume that it is desired to know the capacity of a VVC at −2 volts, if the capacity is 22 picofarads at −4 volts (such as the MV1872 listed in Table 11-1). Since −2 volts is less reverse bias than −4 volts, the capacity will be increased (by a ratio of the two voltages):

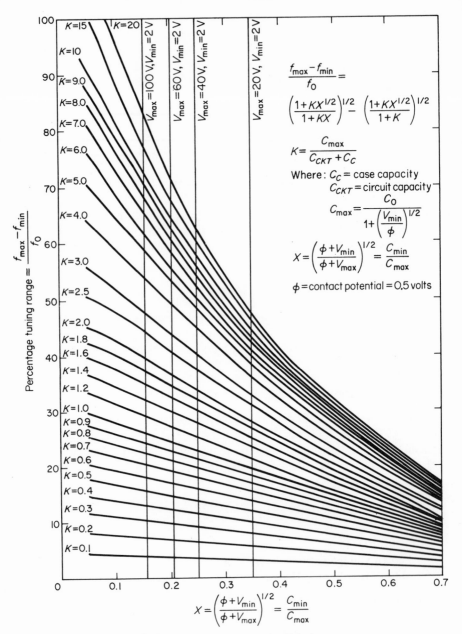

Fig. 11-9. Percentage resonant circuit tuning range.

$$\left(\frac{1 + 4/0.5}{1 + 2/0.5}\right)^{1/2} = 1.34 \text{ ratio} \tag{11-6}$$

$$1.34 \times 22 \text{ pf} = 29.48 \text{ pf} \tag{11-7}$$

The *tuning-range graph* is shown in Fig. 11-9. The following equations form the basis for this graph.

The *resonant frequency* of the circuits of Fig. 11-7 or 11-8 is given by:

$$f_o = \frac{1}{6.28(LC_T)^{1/2}} \tag{11-8}$$

where:

$$C_T = C_{ckt} + C_j + C_c$$

$$C_j = \frac{C_o}{(1 + V/\Phi)^Y} = \frac{C_o}{(1 + 2V)^{1/2}} \tag{11-9}$$

where:

Y is the power law of the junction (0.5 for silicon)
C_o is the zero bias capacity
Φ is the contact potential (0.5 nominal)
V is the magnitude of the reverse voltage

The *maximum capacity* occurs when the minimum bias voltage is applied, and vice versa for the minimum capacity:

$$C_{max} = \frac{C_o}{(1 + V_{min}/0.5)^{1/2}} \tag{11-10}$$

$$C_{min} = \frac{C_o}{(1 + V_{max}/0.5)^{1/2}} \tag{11-11}$$

The *average* or *mean* capacity between C_{max} and C_{min} is usually chosen to resonate the circuit at the design center frequency:

$$C_{cf} = (C_{max}C_{min})^{1/2} \tag{11-12}$$

From the equations 11-10 and 11-11:

$$C_{min} = C_{max}\frac{(0.5 + V_{min})^{1/2}}{(0.5 + V_{max})^{1/2}} \tag{11-13}$$

$$C_{cf} = C_{max}\frac{(0.5 + V_{min})^{1/4}}{(0.5 + V_{max})^{1/4}} \tag{11-14}$$

Defining:

$$\frac{(0.5 + V_{min})^{1/2}}{(0.5 + V_{max})^{1/2}} = X \tag{11-15}$$

$$C_{min} = XC_{max} \tag{11-16}$$

$$C_{cf} = X^{1/2}C_{max} \tag{11-17}$$

The *maximum, minimum,* and *center resonant frequencies* are given by:

$$f_{\max} = \frac{1}{[L(C_{ckt} + C_c + C_{\min})]^{1/2}}$$

$$= [L(C_{ckt} + C_c + XC_{\max})]^{1/2} \tag{11-18}$$

$$f_{\min} = \frac{1}{[L(C_{ckt} + C_c + C_{\max})]^{1/2}} \tag{11-19}$$

$$f_o = \frac{1}{[L(C_{ckt} + C_c + C_{cf})]^{1/2}}$$

$$= \frac{1}{[L(C_{ckt} + C_c + X^{1/2}C_{\max})]^{1/2}} \tag{11-20}$$

Tuning range percentage is:

$$\frac{f_{\max} - f_{\min}}{f_o} = \left[\frac{C_{ckt} + C_c + X^{1/2}C_{\max}}{C_{ckt} + C_c + XC_{\max}}\right]^{1/2}$$

$$= \left[\frac{C_{ckt} + C_c + X^{1/2}C_{\max}}{C_{ckt} + C_c + C_{\max}}\right]^{1/2} \tag{11-21}$$

Defining:

$$K = \frac{C_{\max}}{C_{ckt} + C_c}$$

Equation 11-21 can be simplified to:

$$\frac{f_{\max} - f_{\min}}{f_o} = \left(\frac{1 + KX^{1/2}}{1 + KX}\right)\left(\frac{1 + KX^{1/2}}{1 + K}\right)^{1/2} \tag{11-22}$$

Equation 11-22 is plotted versus $X = C_{\min}/C_{\max}$ for several $K = C_{\max}/(C_{ckt} + C_c)$ as parameters in Fig. 11-9.

Using Fig. 11-9, circuits can be designed and VVCs specified. Actually, Fig. 11-9 can be used for any variable capacitor so long as the abscissa is kept in terms of the minimum to maximum capacity ratio. Using VVCs, the abscissa is also given in terms of minimum and maximum voltage.

Using the tuning range graph. The following is an example of how the graph of Fig. 11-9 can be used.

Assume a tuning range is desired between 60 and 90 MHz, with a fixed circuit capacity of 10 pf present. (This might be due to collector capacity.)

The percentage tuning range is:

$$\frac{90 - 60}{(90 \times 60)^{1/2}} \times 100\% = 41\% \tag{11-23}$$

The maximum usable voltage for any of the VVCs listed in Table 11-1 is 60 volts. That is, the reverse voltage must not be more negative than -60 volts. Therefore, the most negative reverse voltage determines the maximum usable voltage.

The opposite voltage limit (nearest to zero volts) is determined by temperature stability and/or intermodulation effects because the junction capacity varies sharply at low voltages. Temperature dependence enters via the contact potential which is more significant with low applied-bias voltages. It is generally accepted that a lower limit of -2 volts is to be used for most applications.

Therefore, for a V_{max} of -60 volts, a V_{min} of -2 volts, and a tuning range percentage of 41 per cent, the minimum allowable K is 2.5.

Using a C_{ckt} of 10 pf, a C_c of 0.3 (typical for a glass VVC), and the constant K of 2.5, calculate C_{max} as follows:

$$C_{max} = K(C_{ckt} + C_c) \qquad (11\text{-}24)$$
$$\text{or } 2.5(10 + 0.3) = 26 \text{ pf}$$

The C_{max} of 26 pf is for a -2 volt reverse bias; However, since the capacities are listed in Table 11-1 for a -4 volt reverse bias, the C_{max} should be related to a -4 volt capacity. This can be done by calculating the ratio of the two voltages (as shown in equation 11-6), and then dividing the -2 volt C_{max} by this ratio, as follows:

$$26 \text{ pf}/1.34 = 19.4 \text{ pf}$$

Any VVC having a -4 volt capacity greater than 19.4 pf would tune the desired range. However, usually the lower the capacity value the higher the Q so that the lowest capacity device should be selected (where Q is important).

For this problem, the recommended Motorola VVC (in Table 11-1) would be the MV1872 which has a 22 pf -4 volt capacity. This allows for a standard tolerance of ± 10 per cent in the VVC capacity.

Where the best temperature stability is desired, the minimum voltage should be as high as possible. In that case, the X is reduced so that the minimum tolerable K is increased. Larger VVCs (such as the Motorola MV1878 with a -4 volt capacity of 47 pf) would be required. The same condition would occur if the maximum available control voltage were limited.

Assuming that a 22 pf VVC were used (with a ratio of 1.34), and that the circuit capacity was 10 pf, plus a VVC case capacity of 0.3 pf (39.8 pf total), a coil of 0.179 microhenries would be required at 60 MHz. To account for variations of ± 2.2 pf tolerance (22 pf ± 10 per cent), it is recommended that the coil be tunable (by insertion of a ferrite slug, or some similar method).

The *maximum voltage* needed can be calculated from:

$$K = \frac{22 \times 1.34}{10 + 0.3} = 2.85 \qquad (11\text{-}25)$$

From Fig. 11-9 with a 41 per cent tuning range, and a K of 2.85, the required X is 0.24. Using this value,

$$V_{max} = \frac{(2.5)}{(0.24)^2} - 0.5 = 43 \text{ volts} \qquad (11\text{-}26)$$

In the worst case, where the 22 pf value is at the lower limit of 19.8 pf, then the K factor would be:

$$K = \frac{19.8 \times 1.34}{10.3} = 2.55$$

In that event, the full -60 volts would be needed for V_{max}.

For cases where V_{min} and V_{max} have to remain fixed no matter what the circuit and VVC values are (within the prescribed tolerances) the circuit has to be designed around a constant K value. This means $C_{max}/(C_{ckt} + C_c)$ has to be fixed no matter what C_{max} or C_{ckt} values are used. Also, for tuning the desired frequency, $C_{max} + C_{ckt} + C_c$ has to remain constant.

$$L(C_{max} + C_{ckt} + C_c) = \frac{1}{(6.28 f_{min})^2} \tag{11-27}$$

$$\frac{C_{max}}{C_{ckt} + C_c} = K \tag{11-28}$$

By placing a trimmer capacitor in the circuit in addition to a variable inductance, the conditions of equations 11-27 and 11-28 can be maintained.

Using the figures of the previous example, with C_{max} $19.8 \times 1.34 = 26.5$ pf, $f_{min} = 60$ MHz, $K = 2.5$, $C_c = 0.3$ pf.

$$L(C_{ckt} + 26.8 \times 10^{-12}) = \frac{1}{(6.28 \times 62 \times 10^6)^2} \tag{11-29}$$

$$C_{ckt} = \frac{C_{max}}{K} - C_c = \frac{26.5}{2.5} - 0.3 = 10.3 \text{ pf} \tag{11-30}$$

$$L = 0.192 \text{ microhenries}$$

If $C_{max} = 24.2 \times 1.34 = 32.5$ pf, $f_{min} = 60$ MHz, $K = 2.5$, $C_c = 0.3$.

$$C_{ckt} = \frac{32.5}{2.5} - 0.3 = 12.7 \text{ pf} \tag{11-31}$$

$$L = 0.156 \text{ microhenries}$$

Therefore, to maintain constant tuning voltage within the standard VVC tolerances, the circuit inductance must be tunable from 0.156 to 0.192 microhenries, and a trimmer capacity adjustable from 0.3 to 2.7 pf has to be added to the 10-pf circuit capacity. A summary of this design procedure is as follows:

If $f_{max}, f_{min}, C_{ckt}, V_{max}$, and V_{min} are given:

(1) From f_{max} and f_{min} determine $f_o = (f_{max}f_{min})^{1/2}$

(2) Obtain percentage bandwidth $= \dfrac{f_{max} - f_{min}}{f_o}$

(3) Determine $X = \left(\dfrac{0.5 + V_{min}}{0.5 + V_{max}}\right)^{1/2} = \dfrac{C_{min}}{C_{max}}$

(4) Using Fig. 11-9, find $K = C_{max}/(C_{ckt} + C_c)$ at the intersection of constant percentage bandwidth and X lines.

(5) Knowing C_{ckt} (use maximum value possible) and assuming 0.3 for C_c, determine a trial C_{max}, termed C'_{max}.

(6) The VVC selected must have a -4 volt capacity greater than:

$$C_{-4vdc} > 1.1 \times \left(\frac{0.5 + V_{min}}{4.5}\right)^{1/2} = C'_{max}$$

The 1.1 is to allow for the standard 10 per cent VVC tolerance (1.05 for a 5 per cent tolerance).

(7) Once a VVC is selected, assume the minimum possible -4 volt capacity and calculate:

$$C_{max} = \left(\frac{4.5}{0.5 + V_{min}}\right)^{1/2} C_{-4vdc}$$

(a) Using K from step 4, calculate a new C_{ckt}. The difference between this new C_{ckt} and that of the circuit has to be positive. Otherwise, a high capacity VVC has to be selected. The capacity difference is the minimum value required for a trimmer capacitor placed in parallel with the VVC.

(b) Determine the inductance from:

$$L = \frac{1}{(6.28 f_{min})^2 (C_{max} + C_{ckt} + C_c)}$$

where C_{ckt} includes the trimmer value. This inductance L is the *maximum* value of the variable L.

(8) Assume the maximum possible $-4vdc$ capacity, and calculate a corresponding:

$$C_{max} = \left(\frac{4.5}{0.5 + V_{min}}\right)^{1/2} C_{-4vdc}$$

(a) Using K from step 4, calculate a new C_{ckt}. The difference between this new C_{ckt} and that of the circuit is the maximum value required for the trimmer capacity.

(b) Determine inductance as in step 7(b) only, using the C_{ckt} and C_{max} for this step. This inductance L is the *minimum* value of the variable L.

(9) Check the calculations by determining f_{max} using V_{max} for both upper and lower VVC tolerances.

$$C_{min} = C_{max} \left(\frac{0.5 + V_{min}}{0.5 + V_{max}}\right)^{1/2}$$

$$f_{max} = \frac{1}{6.28(C_{min} + C_{ckt} + C_c)^{1/2}(L)^{1/2}}$$

Be sure to use a consistent set of values for C_{min}, C_{ckt}, and L in one case from step 7, and in the other case from step 8.

CHAPTER 12

Typical Semiconductor Circuits

This chapter is devoted to basic circuits that use semiconductor devices described in other parts of this book. No attempt is made to provide a detailed analysis of the circuit. For a comprehensive study of circuits, reference should be made to *Directory of Electronic Circuits* by Matthew Mandl, Prentice-Hall, Inc.

The data in this chapter provide a summary of semiconductor circuits, and serve as a foundation for recognizing and understanding similar circuits found in the semiconductor field.

12-1. Detection or Demodulation Circuits

Detection or demodulation circuits are similar to the rectification circuits discussed in Chapters 4 and 5. However, detection (or demodulation) circuits are used to remove the audio or video information from a modulated r-f carrier.

Basic Diode Detector (AM)

The basic diode-detector circuit for amplitude modulation (AM) is shown in Fig. 12-1.

The modulated AM carrier is transferred (from a previous resonant circuit) by transformer T_1 by a resonant circuit composed of C_1 and the secondary of T_1. A signal voltage is developed across the high impedance.

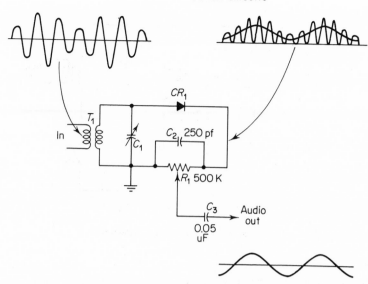

Fig. 12-1. Basic diode detector.

The modulated carrier is a composite signal containing the carrier and sideband components. The semiconductor diode rectifies the modulated carrier. The resultant pulsating d-c is filtered by R_1 and C_2. The amplitude variations of the pulsating d-c are converted to an average d-c whose amplitude changes correspond to the audio-frequency and amplitude used to modulate the carrier.

Such audio signals developed across R_1 are picked off by the variable arm (volume control), and coupled to a subsequent stage by C_3. Capacitor C_3 transfers the audio signals (essentially a-c in character) while keeping the d-c component confined to the detector circuit.

Regenerative Detector

With the regenerative detector circuit of Fig. 12-2, the incoming AM signal is transferred to the resonant circuit by the transformer arrangement of L_1 and L_3. The transistor is operated in a conventional grounded-emitter circuit, and the rectification (detection) occurs between the base and emitter. These two elements act as a conventional diode detector system. The varying signal current between base and emitter also influences collector current flow. As a result, amplification occurs in the emitter-collector circuit.

Collector current is passed through L_2 (known as a *tickler coil*). Coil L_2 feeds back in-phase signals to the base through L_3. Positive feedback is increased by R_1 to the point just below where the circuit would go into self-sustaining oscillation. The feedback signals reinforce the input signals to increase circuit gain.

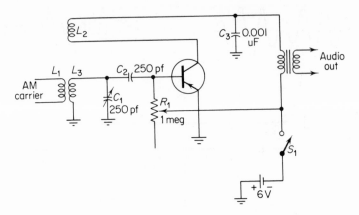

Fig. 12-2. Regenerative detector.

Discriminator Detector (FM)

With the discriminator detector circuit of Fig. 12-3, the FM carrier input (from the i-f amplifier stages) is coupled from L_1 by transformer action to L_2. A portion of the signal of L_1 is also sampled by L_4, and injected into the electrical center of the discriminator output circuit as shown. An alternative method is to derive the sampled portion by eliminating L_4, and coupling this line to L_1 by a series capacitor.

When the incoming carrier is unmodulated, there is a symmetrical circuit balance, and each diode conducts an equal amount of current, on opposite half-cycles. When the FM carrier shifts in frequency, one diode conducts more than the other, and a difference voltage is developed across R_1 and R_2. This difference voltage is the audio component of the FM carrier modulation.

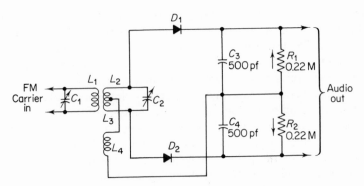

Fig. 12-3. Discriminator detector (FM).

Ratio Detector (FM)

The ratio detector of Fig. 12-4 is similar to the discriminator, except that current flows through R_1 and R_2 in the same direction (because diodes D_1 and D_2 are reversed). The total voltage drop across R_1 and R_2 remains the same, with or without deviation of the FM carrier signal frequency. However, if the carrier deviates from the center frequency (when it is modulated) one diode will conduct more than the other, and the drop across one resistor will be different than the drop across the other resistor. Also, the drop across one resistor will be different in relation to the ground. Thus, the *ratio* of the voltage drops across R_1 and R_2 changes, in proportion to the audio component of the FM carrier modulation. Audio can be taken from either resistor R_1 or R_2, but not across both as with the discriminator.

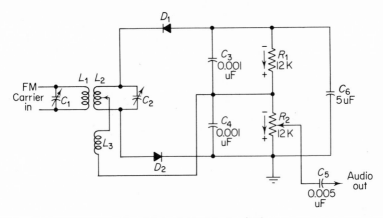

Fig. 12-4. Ratio detector (FM).

Video Detector

A semiconductor diode is often used in TV receivers as the video detector. Such a circuit is shown in Fig. 12-5. Operation of this circuit is the same as that of the basic diode detector (Fig. 12-1). The output voltage is developed across R_1. The capacitors and chokes are used to remove the sound i-f signal.

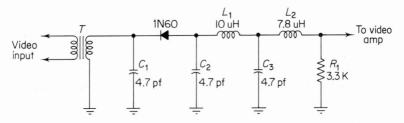

Fig. 12-5. Video detector (TV).

Automatic Volume Control

The basic diode detector is often used to provide an automatic volume control (AVC) function in radio receivers. The circuit shown in Fig. 12-6 is used where it is desirable to maintain a fairly constant audio output level, regardless of the strength of the incoming r-f signal. AVC systems produce a bias level dependent upon the strength of the incoming signal, and apply such a bias to the r-f and i-f stages. If there is an increase in signal strength, the bias is also increased to reduce the r-f and i-f stage gain.

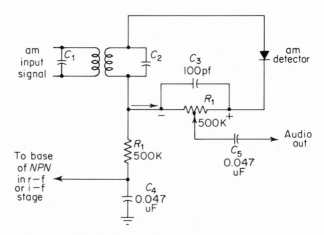

Fig. 12-6. Automatic volume control.

In addition to the basic detector circuit, R_1 and C_4 form a filter section and AVC output network. The voltage at the junction of C_2 and R_1 is always negative with respect to ground. Therefore, the voltage across R_1 is also negative. This negative is applied to the base of an *NPN* transistor in the r-f or i-f stages. When a strong signal is received, the amplitude of the negative AVC voltage increases (more negative), reverse biasing the r-f and i-f stages (or stage), thus reducing the overall output of the radio receiver.

Similar circuits are used to provide automatic gain-control (AGC) functions in television receivers.

12-2. Gating or Logic Circuits

Gating or logic circuits are used primarily in digital work, but are also used in various branches of electronics for passing desired signals, while rejecting undesired signals. Gate circuits are also used to gate-in signals at precise time intervals.

Inverters

Inverter circuits are used when it is necessary to reverse the phase or polarity of a particular signal. A grounded-emitter transistor circuit produces a phase reversal of the signal (negative input/positive output, and vice versa). Semiconductor inverters also often perform an additional function of producing an inverted output, but only in the presence of a positive (or negative) input. Such a circuit is shown in Fig. 12-7.

Note that there is no forward bias applied to the base-emitter, but reverse bias is applied to the emitter-base. No current flows without an input signal. When a negative pulse is applied to the base, the emitter-base junction is forward biased, and current flows in the emitter-collector circuit, for the duration of the input pulse. The circuit output is an amplified, phase-inverted version of the input signal.

In digital applications, such phase inverter circuits are sometimes referred to as NOT circuits. This term is derived from the fact that the phase of the output signal is "not" the same as that of the input signal.

Field Effect Transistors (FET) are being used to simplify logic circuits. For switching and logic circuits, complementary enhancement type FETs offer two significant advantages. Power is drawn only during switching, not in either stable state, and no coupling elements are required since the input to each FET resembles a capacitor and the coupling is inherent in the device.

One simple circuit that illustrates these advantages is the complementary inverter of Fig. 12-8. In this circuit, $+V$ is a logical "1" and ground is a logical "0." If the input signal is $+V$, the P-channel device (upper) is essentially off. The N-channel FET is forward biased but since only I_{DSS} is available from the upper stage, V_{DS} is very low. As a result, the potential of the output is near ground—the zero reading. When the input goes to ground, the P-channel FET conducts. The N-channel device, however, is now cut off and conducts only I_{DSS}. Since the voltage drop across the P-channel device

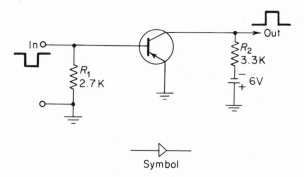

Fig. 12-7. Basic transistor inverter.

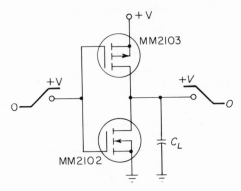

Fig. 12-8. FET complementary inverter circuit *(courtesy Motorola).*

is very low, the drain potential of the P-channel device is approximately $+V$. This charges capacitor C_L to $+V$.

OR or NOR Circuits

An OR circuit produces an output pulse whenever a signal is applied to one of the input terminals, or all of the input terminals simultaneously.

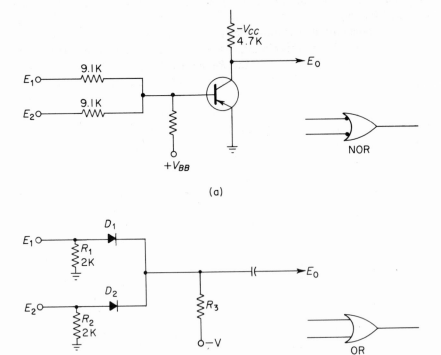

(a)

(b)

Fig. 12-9. OR or NOR circuits.

A transistorized OR circuit is shown in Fig. 12-9(a). A negative input signal at either E_1 or E_2, or both, produces an inverted (positive) output. Since there is a phase reversal between input and output, this OR circuit also has NOT characteristics. For this reason, such a circuit is sometimes known as a NOT-OR circuit or a NOR circuit.

An OR circuit using diodes is shown in Fig. 12-9(b). Both diodes are forward biased by the battery voltage. A positive pulse at either input (or both inputs) will increase the forward bias, and produce a positive pulse at the output. Since there is no phase reversal across this OR circuit, it has no NOT characteristics, and can not be used as a NOR circuit.

AND and NAND Circuits

With AND circuits (also called *coincidence gates*) a signal must be present at both inputs simultaneously to produce an output signal.

A transistorized AND circuit is shown in Fig. 12-10(a). A negative input signal at either E_1 or E_2, but *not at both terminals*, will forward bias the corresponding transistor. However, because the opposite transistor is in a

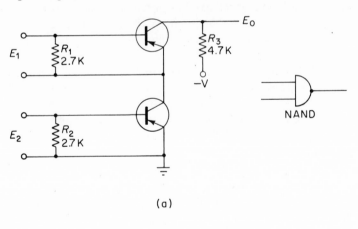

(a)

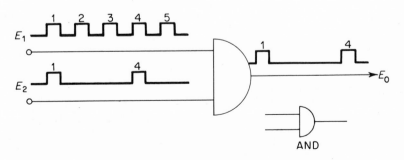

(b)

Fig. 12-10. NAND circuit (coincidence gate).

nonconducting state, no output signal will be produced. When a negative signal is applied to both inputs, the coincidence of the signals will permit both transistors to conduct since the forward bias is present for both. The output signal will have a positive polarity, producing a NOT function. Such a circuit is sometimes referred to as a NOT-AND circuit or a NAND circuit.

The function of the AND circuit is shown in Fig. 12-10(b). Here, four pulses are applied to the upper input terminal as shown, while only two pulses are applied to the lower input terminal. Since the latter two pulses coincide with the first and fourth pulses applied to the upper input terminal, coincidence prevails only for these two conditions. Consequently, the output pulse waveform would consist only of the first and fourth pulses as shown. Thus, the circuit exhibits gating-in characteristics because the lower pulses essentially gate-in any pulses appearing simultaneously at the upper input terminal.

Inhibiting Circuit

An inhibit circuit is one which exhibits *gating-out* characteristics, and is essentially the opposite of an AND circuit. With an inhibit circuit, there is an output only when one input is present, but never when two inputs are present.

A transistorized inhibit circuit is shown in Fig. 12-11(a). If a negative pulse is applied to Input 1 (with nothing at Input 2), the pulse will be inverted by the transformer, and a positive pulse will be applied to the transistor base. This will forward bias the transistor, and an output will appear.

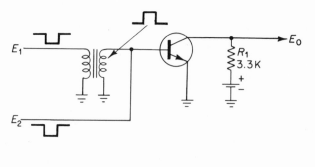

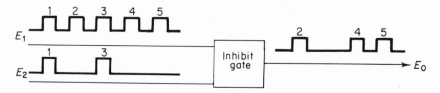

Fig. 12-11. Inhibit gate.

If a negative pulse is applied to Input 2, the transistor will be driven further reverse-biased, and no output will appear.

If a negative-pulse is applied to Input 2, simultaneously with a pulse at Input 1, the two inputs will be cancelled at the base, and the transistor will remain nonconducting.

Therefore, an output will appear only when a pulse is applied to Input 1, with no pulse at Input 2.

The function of the inhibit circuit is shown in Fig. 12-11(b). Here, a series of pulses is applied to the upper terminal, but only two pulses appear at the lower terminal.

Since the lower terminal pulses appear in coincidence with the first and third upper terminal pulses, the lower terminal pulses will inhibit the entry of the upper terminal pulses into the circuit, and prevent their appearance at the output. Consequently, only the second, fourth, and fifth pulses appear at the output as shown. Thus, such a circuit can be considered as a gating-out circuit since the lower pulses function to gate-out the appearance of the upper pulses from the output pulse train.

12-3. Amplifier Circuits

The function of any electronic amplifier is to increase the voltage or power amplitude of the input signal. Since a transistor is essentially a current-operated device, transistor amplifiers are current (or power) amplifiers. Except for this basic characteristic, and some minor differences discussed in later paragraphs, transistor amplifiers are essentially the same as the equivalent vacuum tube circuits.

Transistors can be operated in any type of amplifier (audio, i-f, r-f) and at any class (*A*, *B*, *C*, or intermediate classes). It is assumed that the reader is familiar with the basic theory of amplifiers and amplifier classes, so such data will not be repeated here. Instead, the transistor version of typical amplifier circuits is given.

Class A Audio Amplifier

Figure 12-12 shows the circuit for a typical Class *A* audio amplifier. This circuit is suitable for most audio applications, except where high power is required. A common-emitter configuration is used. Forward bias for the base-emitter is provided through R_1, while reverse bias for the base-collector is obtained through R_2. Resistor R_2 also functions as a load resistor, to develop an output signal (by voltage drop across R_2) in response to emitter-collector current variations.

Resistor R_1 provides a forward bias value that will cause emitter-collector current to flow under no-signal conditions, as well as maximum input signal conditions. Since current flows at all times, the circuit is operated as Class *A*.

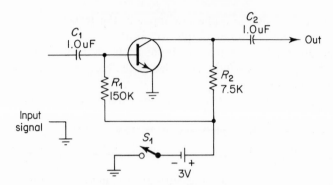

Fig. 12-12. Class A audio amplifier.

Direct Coupled Amplifier

Transistors can be used to great advantage in direct coupled amplifiers, where the collector output of one transistor is fed directly into the base input of the succeeding stage, without a coupling capacitor. The use of complementing transistors permits design of a direct coupled amplifier without the disadvantage of two voltage supplies.

Figure 12-13 shows a typical direct coupled circuit using an *NPN* and a *PNP* transistor. The forward and reverse bias potentials of both transistors are supplied by a single power source.

The *NPN* emitter is connected to the negative battery terminal through the ground lead. The positive battery terminal is connected across R_1 and R_2 which act as voltage dividers. A positive potential appears at the *NPN* base, making the base positive with respect to the emitter, and producing forward bias. The *NPN* collector is positive with respect to both the base (for reverse bias) and emitter, since the collector is connected to the positive battery terminal through R_3.

The *PNP* emitter is connected to the positive terminal of the battery for

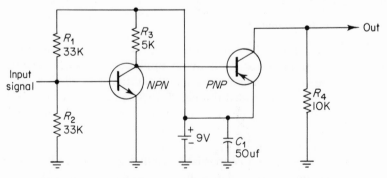

Fig. 12-13. Direct coupled amplifier (complementing transistors).

forward bias. The *PNP* base is connected to the *NPN* collector. Although this point is positive with respect to ground, it is negative with respect to the *PNP* emitter (by the voltage drop across R_3). Therefore, the *PNP* base-emitter is forward biased. Resistor R_4 supplies the necessary negative potential for the *PNP* collector.

Audio Power Amplifier

An audio power amplifier converts the signal voltage (or current) into an equivalent signal *power* for application to a loudspeaker or other load, as well as to drive other amplifiers where the input system requires signal power. A typical transistorized audio power amplifier is shown in Fig. 12-14. This circuit is sometimes referred to as a single-ended amplifier.

The interstage transformer is used to step down the high impedance of the Q_1 collector output circuit to the relatively low input impedance of Q_2. Signal voltages developed across the transformer secondary, alternately add and subtract from the forward bias between the base and emitter of Q_2, by aiding or opposing the fixed bias amplitude. Thus, current changes occur in the collector side of Q_2 and represent signal power. An audio output transformer is used to provide an impedance match between the loudspeaker voice coil and the recommended load resistance for Q_2.

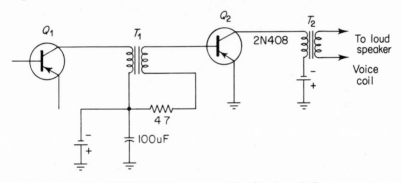

Fig. 12-14. Audio power amplifier (single ended).

Push-Pull Audio Amplifier

Figure 12-15 is a transistorized push-pull audio amplifier. The emitter forward bias for Q_2 is obtained through the loudspeaker voice coil, while the collector reverse bias is obtained by the common-ground connection. The positive for the base of Q_2 is taken from voltage divider resistors R_1 and R_2 (shunted across battery B_2).

Resistors R_3 and R_4 shunt battery B_1, and provide forward bias for Q_3. The positive potential for reverse bias of the Q_3 collector is applied through the loudspeaker voice-coil inductance.

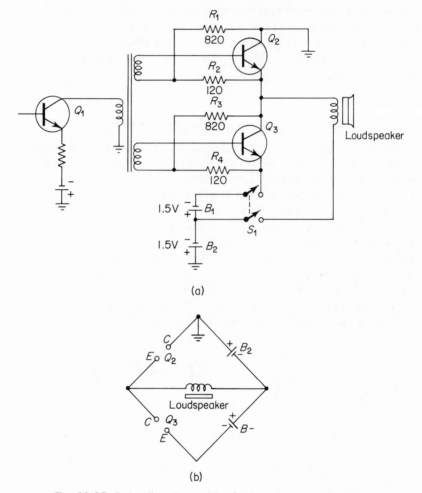

Fig. 12-15. Push-pull audio amplifier (without output transformer).

The interstage coupling transformer has a split secondary to provide the required out-of-phase signal voltages for the push-pull inputs. Each transistor base has a signal applied to it which is out-of-phase with that of the other transistor base.

The two transistors form a bridge as shown in Fig. 12-15(b). On alternate half-cycles of the audio input signal, one transformer conducts more than the other, and an unequal current flow occurs, unbalancing the bridge. The unbalance causes audio signal voltages to develop across the loudspeaker voice coil.

With a transistor push-pull circuit, the loudspeaker load impedance required for matching is only a fraction of that normally needed in a vacuum tube circuit.

Class A R-F Amplifier

Class A r-f amplifiers are used as the input circuit of some radio communication receivers. A typical circuit is shown in Fig. 12-16. This circuit is the equivalent of a vacuum tube Class A r-f amplifier, with two exceptions. The forward bias for the base-emitter section is obtained from an AVC voltage. (AVC is discussed in Sec. 12-1.) Changes in AVC voltage regulate the stage gain in proportion to the level required to maintain a fairly constant volume for input signals of differing strength.

The major difference in the circuit of Fig. 12-16 from a vacuum tube circuit is the RC network C_4-R_2. This network provides increased thermal stability, but does not supply bias (as would be the case for a cathode resistor in vacuum tube circuits). Current flow through R_2 causes a voltage drop to occur which has a polarity opposite to the battery voltage. This reduces the 9-volt supply by a value dependent upon the voltage drop across R_2. If a temperature change tends to increase current through the transistor (a temperature increase), the voltage drop across R_2 increases causing a net reduction in collector-emitter voltage and a compensating decrease in transistor current. Capacitor C_4 provides a bypass effect and minimizes voltage variations across R_2.

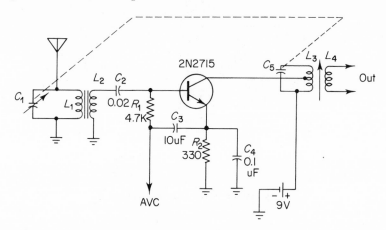

Fig. 12-16. Class A r-f amplifier.

Grounded Base Amplifier

The grounded base circuit is often used in high frequency r-f amplifiers. (Fig. 12-17.) As frequency increases, the capacitive coupling between transistor elements may produce some feedback and possible regeneration between input and output. This can result in oscillation unless the circuit is neutralized.

With a grounded base, an effective electrostatic shield exists between the input and output circuits, and the tendency to oscillate is at a minimum.

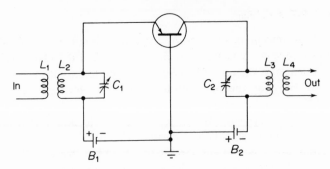

Fig. 12-17. Grounded base amplifier.

Class B Audio Amplifier

Figure 12-18 shows a Class *B* audio amplifier. Except for the bias, the circuit is the same as that of an equivalent vacuum tube amplifier. Bias for the circuit of Fig. 12-18 is obtained from voltage divider resistors R_1 and R_2. Each transistor conducts on alternate half-cycles of the input signal, and is biased at cut-off (no collector current flow) during the opposite half-cycle.

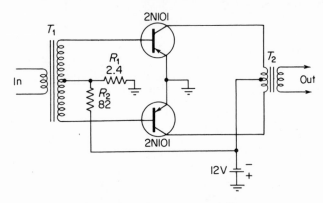

Fig. 12-18. Class *B* audio amplifier.

Class A I-F Amplifier

The Class *A* i-f amplifier of Fig. 12-19 is similar to the Class *A* r-f amplifier of Fig. 12-16, except for the method of tuning and the neutralization capacitor C_3. Again, emitter resistor R_2 and its bypass C_5 are for thermal stability, not bias.

Class C R-F Amplifier

Class *C* r-f amplifiers are used as the output or multiplier circuits of some radio communication transmitters. A typical circuit is shown in Fig. 12-20.

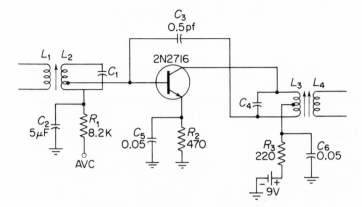

Fig. 12-19. Class A i-f amplifier.

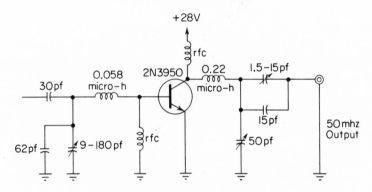

Fig. 12-20. Class C r-f amplifier.

Note that the base-emitter circuit does not require forward bias, since the circuit is to operate only in the presence of an input signal. However, the base-collector circuit is reverse biased by the 28 volts applied to the collector (and ground). The emitter-collector circuit is completed through the r-f choke in the collector line.

Since the transistor is an *NPN*, the base-emitter circuit will be forward biased each time the input signal swings positive (at the base). This will cause current to flow in the emitter-collector circuit.

Emitter Follower

Although not classified as an amplifier, it is possible to obtain *power gain* from an emitter-follower circuit shown in Fig. 12-21. This is because the current through the emitter-collector circuit (through R_2) can be many times that of the input signal current.

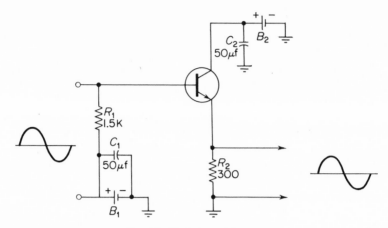

Fig. 12-21. Emitter follower.

12-4. Oscillator Circuits

There are two basic types of transistor oscillator circuits. The *resonant* circuit oscillators produce a signal frequency related to the resonance established by an inductor-capacitor circuit. The *relaxation* circuit generates a signal whose frequency is related to the value of circuit components (usually *RC* components).

Basic Oscillator Circuits

Figure 12-22 shows the circuits of some typical transistor resonant circuit oscillators, while relaxation oscillators are shown in Fig. 12-23. Each of these circuits is the counterpart of a corresponding vacuum tube circuit.

Unijunction Oscillator

The unijunction oscillator is of the relaxation type, but is unique to semiconductor circuits. A typical unijunction oscillator is shown in Fig. 12-24.

The positive supply voltage causes C_1 to charge at a rate depending upon the time constants of R_1, R_2, and C_1. When the charge reaches a level high enough to trigger the unijunction, the impedance between the emitter and base 1 is lowered. C_1 then discharges. This drops the emitter into the reverse-bias region, stopping conduction. Capacitor C_1 charges again, and the cycle is repeated. The oscillation frequency is dependent upon the R_1, R_2, and C_1 time constants.

A synchronization signal can be applied across resistor R_2 to lock in the oscillator, if desired. Oscillator output is taken from the voltage drop

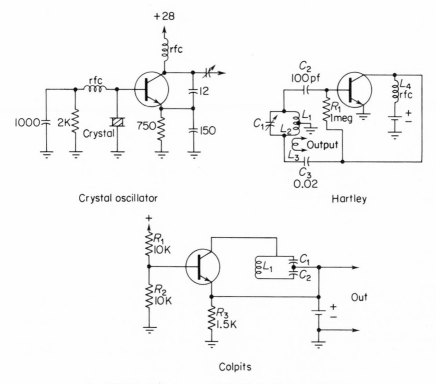

Crystal oscillator   Hartley

Colpits

Fig. 12-22. Typical resonant circuit transistor oscillators.

across resistor R_4. A secondary saw-tooth signal can also be taken from the emitter. This sawtooth is the result of capacitor C_1 charge and discharge.

12-5. Power-Control Circuits

SCRs, SCSs, *PNPN* switches, Triacs, and similar forms of thyristors are often used in power-control circuits. These devices are particularly useful where it is necessary to control a large amount of power with a small signal voltage. In some cases, control may only involve the simple turning on or off of the power supplied. On other applications, it is necessary to undertake precise control of the amount of power furnished.

Basic Power-Control Circuit

The basic power-control circuit using a controlled rectifier is shown in Fig. 12-25. Here, d-c power to the load is controlled by an SCR (or SCS, or *PNPN* switch). The trigger signal for the SCR is obtained from a phase-shift network.

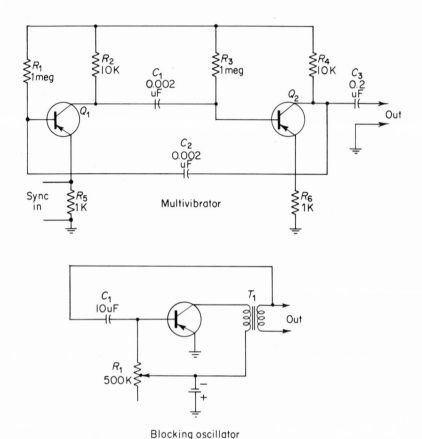

Fig. 12-23. Typical relaxation circuit transistor oscillators.

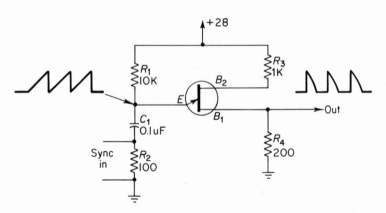

Fig. 12-24. Unijunction oscillator.

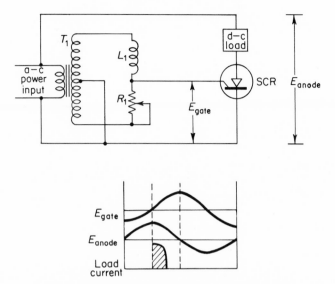

Fig. 12-25. Basic d-c power-control circuit (with SCR).

The phase-shift circuit produces an output which is different in phase from the input a-c. The transformer secondary has a center-tap to provide a 180° phase difference between the upper and lower terminals. An additional inductor L_1 in series with a variable resistor R_1 shunts the secondary winding and provides a means for varying the phase produced. Since the SCR anode receives its voltage from the same a-c line which feeds the transformer primary, the *relative* phase of the signal applied to the SCR trigger can be controlled by R_1.

Figure 12-25 also shows the effect of varying the relative phase between the anode and trigger voltages. Power is applied to the load only when the SCR conducts (indicated by the shaded portion of the load current waveform).

Basic A-C Power-Control Circuit

The circuit shown in Fig. 12-25 is suitable for control of power in d-c circuits, since the SCR rectifies power to the load. As discussed in Chapter 9, a Traic or similar device is better suited to control a-c power circuits.

The basic Traic circuit is shown in Fig. 12-26. Here, the Traic conduction angle (length of time the Triac is turned on during each half-cycle) is determined by the time constants of C_1 and R_1. During each half-cycle, C_1 charges through R_1. A large value of R_1 will cause a slow charge; a lower value of R_1 will increase the charge speed. When the voltage at C_1 reaches the breakdown voltage of the Diac, a trigger will be applied to the Triac gate input. The Triac will then conduct for the remainder of the half-cycle. Therefore,

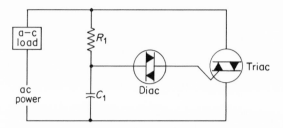

Fig. 12-26. Basic a-c power-control circuit (with Triac).

the conduction angle and the amount of power to the load are set by adjustment of R_1.

12-6. Pulse Shaping Circuits

Semiconductors (particularly diodes) are used frequently as pulse shapers to clip, slice, limit, and clamp pulses and other signals.

Series Diode Clippers

A clipper circuit clips a particular waveform so that the amplitude of the output signal is proportional to those values of input waveform exceeding a certain critical value. A typical diode clipper is shown in Fig. 12-27.

The diode is normally reverse biased by the battery voltage, and no current flows in the absence of an input signal. If a positive input is applied, current will not flow until the input positive amplitude rises above the battery bias voltage (4.5 volts). The output amplitude is then proportional to the amplitude of the input signal which exceeds the 4.5 bias potential value. No current flows for a negative input signal. Therefore, the circuit clips those portions of the signal which are below the positive 4.5-volt bias value.

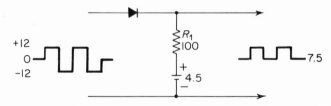

Fig. 12-27. Series diode clipper.

Shunt Diode Clippers

Figure 12-28 shows the four basic types of shunt-clippers. In all four circuits, the diode will be forward biased during a portion of the input waveform and will conduct. During conduction, the low forward resistance of the diode

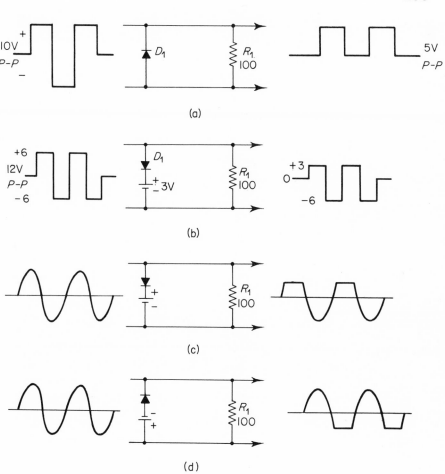

Fig. 12-28. Shunt diode clippers.

shunts the load resistance (R_1), and no output is produced. Thus, that portion of the input waveform is clipped. The diode is reverse biased during the remainder of the input waveform, so the remainder of the waveform is reproduced across the output load resistor R_1.

In (b), (c), and (d) of Fig. 12-28, the diode is reverse biased by the battery voltage. When the input signal exceeds this bias, the diode will conduct, and no output voltage will be developed across R_1. Thus, that portion of the input signal above the bias voltage is clipped. With negative input signals, the diode does not conduct, and the output is proportional to the input.

Parallel Clippers or Slicers

Two shunt clippers can be combined to produce a parallel clipper, or slicer, as shown in Fig. 12-29(a). The same effect can be produced by an over-

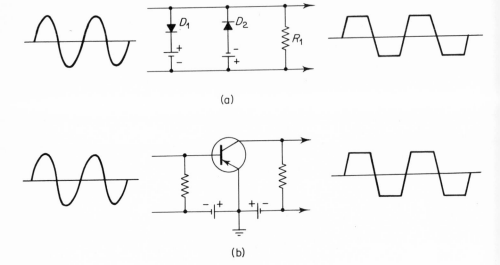

(a)

(b)

Fig. 12-29. Parallel clippers (slicers).

driven, common-emitter, Class *A* transistor amplifier as shown in Fig. 12-29(b).

In either circuit, the purpose is to clip the peaks of an a-c input signal so that the output resembles a squarewave, or pulse. The amount of clipping depends upon the bias voltages in relation to the input signal.

In Fig. 12-29(a), diode D_1 conducts on positive half-cycles while D_2 conducts on negative half-cycles. No output occurs during conduction of either diode, so the peaks are clipped. The greater the bias voltage, the lower the output amplitude.

In Fig. 12-29(b), base-emitter bias must be of a value that will permit the input signal to drive the transistor into saturation, and into cutoff, on positive and negative peaks.

When the transistor is driven into saturation, a further increase in input signal will cause no further increase in output. When in cutoff, a further input increase will produce no further change in output. Therefore, both the positive and negative peaks of the input waveform are clipped (or sliced).

Shunt Limiter

The biased shunt-clipper circuit can be used as a limiter circuit as shown in Fig. 12-30. Here, all input signals which exceed the bias voltage will be limited to the value of the bias voltage. The diode is reverse biased by the battery, and will not conduct on any input pulses equal to, or below, the bias value. Any input pulses that exceed the bias value will cause the diode

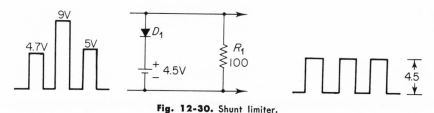

Fig. 12-30. Shunt limiter.

to conduct, and remove the output. Such circuits are often used to remove transients such as overshoots. The circuit will have no effect on pulses of amplitudes below that of the bias voltage.

Clamp or D-C Restorer Circuits

When pulses are passed through capacitor coupling between stages, the d-c component of the pulses is lost. In the circuit of Fig. 12-31, if the clamping diode were not present, the pulses on the output side of capacitor C_1 would be of the same amplitude as the input pulses, except that the output pulses would be positive and negative with respect to ground. For example, assume that the input pulses had a 50-volt peak value. Also assume that the steady-state value on the input side of C_1 was $+25$ volts (at the collector of Q_1). Between pulses, C_1 would charge to the steady-state value. The input side of C_1 would be positive with respect to ground, while the output side would be negative with respect to ground.

When the pulses were present, C_1 would charge up to the 50-volt peak value. On the input side, C_1 goes from $+25$ to $+75$ volts. On the output side, C_1 goes from -25 to $+25$. Therefore, the output pulses are positive and negative with respect to ground, even though the input pulses are always positive with respect to ground.

Diode D_1 prevents this condition by discharging C_1 between pulses. This makes the output side of C_1 zero with respect to ground until a pulse arrives. Then the output side rises to the peak value (in this case, 50 volts).

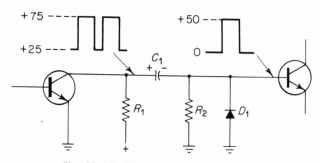

Fig. 12-31. Clamp or d-c restorer circuit.

12-7. Photocell Circuits

There are many uses for photocells, and hundreds of circuits. However, the basic function of the photocell is to produce (in the case of photovoltaic cells) or control (in the case of photoconductive cells) electrical current in proportion to light. The basic circuit is shown in Fig. 12-32.

In Fig. 12-32(a), the output of a photovoltaic cell is used to control operation of a relay through a basic transistor amplifier. The transistor is *PNP*, so the photovoltaic cell output will forward bias the base-emitter circuit. If an *NPN* were used, the cell terminals would be reversed.

When light is applied to the cell, the base-emitter circuit is forward biased, and emitter-collector current flows. If the light is sufficient, the forward bias will cause enough collector current to actuate the relay. As light is reduced, the forward bias will also decrease, lowering the collector current and de-actuating the relay.

This basic circuit could be used to provide automatic control of house lights or an electric sign. At nightfall, light striking the photocell would be reduced so that the relay would de-actuate. The relay contacts could be arranged so that de-actuation would turn on the lights. At dawn, as light striking the cell increased, the relay would be actuated, removing power to the lights.

The circuit of Fig. 12-32(b) is the photoconductive version. Here, the

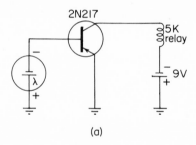

(a)

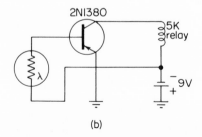

(b)

Fig. 12-32. Photocell circuits.

amount of base-emitter forward bias is controlled (rather than produced) by the light striking the cell. However, operation of the circuit is the same.

12-8. Special Circuits

There are literally thousands of semiconductor circuits that could be considered as "special." Many of these circuits are special in one field of electronics, but commonplace in another. The following paragraphs discuss two circuits which are special versions of other circuits, but which have found widespread use in semiconductor electronics.

Flip-Flop Circuit

The flip-flop circuit (also known as the Eccles-Jordan circuit) of Fig. 12-33 is similar to the multivibrator of Fig. 12-23. However, the flip-flop is not free-running, and is not an oscillator. Instead, the flip-flop must be triggered into each of its alternate states (one transistor in saturation, the opposite transistor cutoff). This makes the flip-flop an excellent divider or counter. The transistor version (or integrated circuit version) of the flip-flop operates in a manner similar to that of the vacuum tube counterpart.

When power is first applied, one transistor is driven to saturation, with its output driving the opposite transistor into cutoff. Assume that Q_1 is saturated and Q_2 is cutoff. A positive trigger pulse across R_5 will cause a reverse bias to be applied to both bases. This pulse will have no effect on

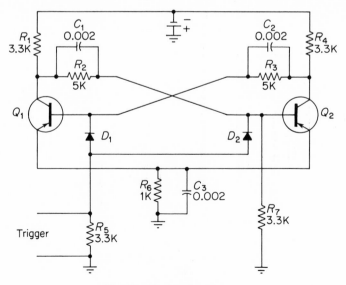

Fig. 12-33. Flip-flop circuit.

Q_2 (already reverse biased), but will reduce the collector current of Q_1. In turn, this reduces the reverse bias on Q_2, increasing its collector current. The condition continues, until the circuit has reached its opposite bi-stable state (Q_1 cutoff and Q_2 saturated).

The output is taken from the collector of Q_2. One trigger pulse will cause the output pulse to go from negative (cutoff) to positive (saturation). The next pulse will reverse the condition, and change the output from a positive (saturation) to a negative (cutoff). Therefore, it takes two trigger pulses to produce one complete output cycle (negative to positive, and back to negative again). For this reason, the flip-flop circuit is often used as a two-to-one divider or counter.

Schmitt Trigger

The Schmitt trigger is another bi-stable circuit. However, the Schmitt circuit returns to its steady-state in the absence of a trigger signal. The Schmitt trigger shown in Fig. 12-34 will provide an output in the presence of an input, of *sufficient amplitude*.

With no trigger input, capacitor C_2 charges to the value of the Q_2 collector. Transistor Q_2 is forward biased by the positive voltage at the junction of R_2 and R_3. Transistor Q_1 is reverse biased by the voltage developed across R_5 (which makes the *NPN* emitter more positive than the base).

With a trigger of sufficient amplitude applied, the reverse bias on Q_1 is overcome, and Q_1 conducts. This drops the voltage on the collector of Q_1, and the base of Q_2, removing the forward bias from Q_2. Transistor Q_2 stops conducting, and an output pulse is produced at the Q_2 collector.

The Schmitt trigger is useful in computer work where it is often necessary to reform and reshape pulses.

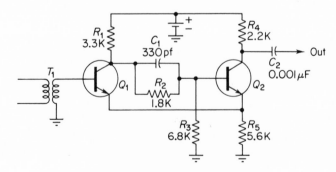

Fig. 12-34. Schmitt trigger.

CHAPTER **13**

Transistor Parameter Data

In Chapter 3, transistor characteristics and test procedures are described from the practical viewpoint. This chapter supplements the practical procedures by providing a summary of transistor operating characteristics (or parameters) from the theoretical approach. While a technician is not usually called upon to solve design problems with the aid of transistor parameters, a senior laboratory technician must be familiar with the nomenclature of all commonly used transistor parameters in order to interpret a manufacturer's data sheets properly.

Transistor parameters are vaguely similar to vacuum tube parameters, in that they both show the relationship of output to input under various conditions. However, transistor parameters are far more complex. The major reason for this is that transistor input-output factors depend upon one another, even when output terminals are open or short circuited to a-c signals. In effect, transistor input and output circuits are *interconnected resistance networks*.

Current and voltages at transistor inputs and outputs are different under open or short-circuited conditions because there is interaction between circuit elements under both conditions, and feedback of energy from output to input occurs when signals are passing through. This feedback produces what is known as *transfer circuit constants* under both open and short-circuited conditions of the input and output terminals. These parameters are called forward and reverse transfer resistance, forward and reverse conductance or admittance, forward and reverse amplification factor, etc.

As a result, two entire sets of open and short-circuit voltage, current and impedance relationships exist (one for input and one for output).

To further confuse matters, each manufacturer has his own set of letters and subscripts to describe the open and short-circuit parameters. In many cases, the parameters are mixed on the data sheets. That is, a single data sheet will contain combinations of parameters taken from more than one parameter system.

To reduce confusion in interpreting transistor data sheets, the technician must:

(1) have an understanding of the commonly used parameters,
(2) have a means of converting from one parameter to another, and
(3) have a means of determining parameters at different frequencies, temperatures, etc., when only one condition is specified on the data sheet.

13-1. Understanding Commonly Used Transistor Parameters

There are a number of systems used to define and measure transistor parameters. The most commonly used include: r or resistance, y or admittance, z or impedance, s or scattering, g or transconductance, and h or hybrid.

The r or *resistance* parameters are not in common use today. This is because r parameters do not take into account any gain between input and output. However, r parameters do provide a foundation for understanding other parameter systems.

A transistor could be considered as a three-terminal resistor network as shown in Fig. 13-1. A T-network is formed by emitter resistance r_e, collector resistance r_c, and base resistance r_b. Since one of the terminals (the base terminal in this case) is common to both input and output, a more practical representation would be the four-terminal networks of Fig. 13-2.

The networks of Fig. 13-1 are passive, and the resistance values represented

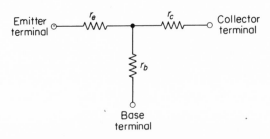

Fig. 13-1. Basic three-terminal resistance network equivalent of a transistor.

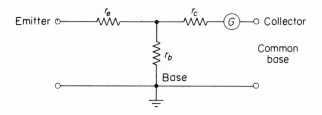

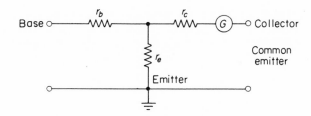

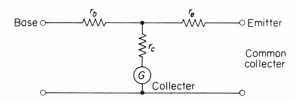

Fig. 13-2. Equivalent transistor networks for d-c and low frequency operation.

would be determined by d-c measurement. The networks of Fig. 13-2 are active in that they show a generator (*G*) in the output to indicate the amplifying function of a transistor. The networks of Fig. 13-2 are a close approximation of true transistor characteristics at low frequencies, and at d-c, but not at high frequencies.

As frequency increases, the internal capacitance of transistors produces impedance. Also, the external circuits in which the transistors are used affect the network. For this reason, it is more convenient to consider transistor networks as "black boxes" similar to that shown in Fig. 13-3. To analyze the black box, the voltages and currents at input and/or output are measured, or signals are applied to input/output, and the signal effect is evaluated.

No matter what parameter system is used, I_1 indicates input current, V_1 indicates input voltage, I_2 indicates output current, V_2 indicates output voltage.

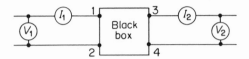

Fig. 13-3. Equivalent black-box representation of transistor parameters.

In considering r parameters, the concern is with the effects of resistance (or voltage divided by current) since $r = E/I$.

For example, r_{11} indicates input resistance, obtained by using values of V_1 and I_1, with the output open circuited. The first 1 of the subscript indicates V_1 voltage, and the second 1 of the subscript indicates the I_1 current.

R_{21} indicates forward transfer resistance, obtained using values of V_2 and I_1. Here, a test voltage V_1 is applied to the input. The current I_1 and the output voltage V_2 are measured.

R_{12} indicates reverse transfer resistance, or V_1/I_2.

r_{22} indicates output resistance, or V_2/I_2.

It should be noted that the double number subscript implies that the first variable is always divided by the second.

The h or *hybrid system* of parameters is the most popular method to indicate transistor characteristics (at least this is true on data sheets). Where r parameters were measured with open circuit conditions, h or *hybrid* parameters are based on combinations of constant current and constant voltage, in both open and short-circuit conditions.

Using the black box of Fig. 13-3, the four basic h parameters are:

$h_{11} = V_1/I_1$ (input resistance, output terminals 3 and 4 shorted, $V_2 = 0$)

$h_{12} = V_1/V_2$ (reverse voltage ratio, input terminals 1 and 2 open, $I_1 = 0$)

$h_{21} = I_2/I_1$ (forward current ratio, output terminals 3 and 4 shorted, $V_2 = 0$)

$h_{22} = I_2/V_2$ (output conduction, input terminals 1 and 2 open, $I_2 = 0$)

The use of number subscripts for h parameters has been replaced (in most practical applications) by letter subscripts.

The first subscript indicates the characteristic: i for input, o for output, f for forward transfer, and r for reverse transfer.

The second subscript indicates the circuit configuration: b for common base, c for common collector, and e for common emitter.

Using this system, h_{11} or input resistance, could be indicated by h_{ib} for common base input, h_{ic} for common collector, and h_{ie} for common emitter.

h_{12} or reverse transfer voltage ratio could be indicated by h_{rb}, h_{rc} and h_{re} for common base, collector and emitter, respectively.

h_{22} or output conductance could be indicated by h_{ob}, h_{oc} and h_{oe} for common base, collector and emitter, respectively.

h_{21} or forward current transfer ratio (usually the most important parameter) could be indicated by h_{fb}, h_{fc} and h_{fe} for common base, collector and emitter, respectively.

When h parameters were first used, the common base configuration was the most popular. Today, the common emitter configuration is used to the greatest extent. For this reason, practically all transistor data sheets list h_{fe}. Some data sheets mix the configurations. For example, the data sheet for General Electric 2N337 lists h_{fe}, h_{ib}, h_{rb}, and h_{ob}.

Practical Hybrid Parameter Applications

Figure 13-4 shows the basic grounded-emitter circuit used for most h parameters, together with the hybrid equivalent circuit.

Figure 13-4 also shows the equations for the most important transistor circuit characteristics: A_i or signal current gain, R_i or input resistance, A_e or A_v signal voltage gain, and A_p or signal power gain. Note that these equa-

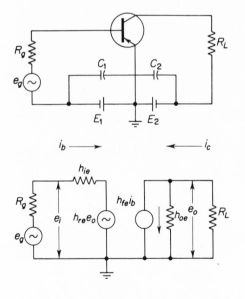

h_{oe} = Output admittance or conductance
h_{ie} = Emitter input resistance
h_{re} = Reverse voltage transfer ratio
h_{fe} = Forward current ratio

$$A_i = h_{fe} / 1 + h_{oe}R_L = \text{Signal current gain}$$

$$R_i = h_{ie} - \frac{h_{fe}h_{re}R_L}{1 + h_{oe}R_L} = \text{Input resistance}$$

$$A_e = \frac{e_o}{e_i} = \frac{1}{h_{re} - \frac{h_{ie}}{R_L}\frac{1+h_{oe}R_L}{h_{fe}}} = \text{Voltage amplification}$$

$$A_p = A_e A_i$$

Fig. 13-4. Grounded emitter and equivalent hybrid circuit.

tions use the letter subscripts, rather than number subscripts. This matches most data sheets for transistors.

The equations of Fig. 13-4 differ from those of Sec. 13-3 which follows. However, the net results are the same. The equations of Fig. 13-4 are based on a common emitter circuit. The same equations apply to common base or common collector circuits. However, the second subscript e must be changed to b or c, as applicable.

The following is an example of how hybrid parameters can be applied to a practical problem. Assume that it is desired to find signal current gain A_i for a GE2N337 transistor used in a circuit with a 10,000-ohm load resistor. The data sheet for a 2N337 shows:

$$h_{fe} = 55 \text{ (typical)}$$

$$h_{ob} = 0.1 \text{ micromho (typical)}$$

The equation is:

$$A_i = \frac{h_{fe}}{1} + h_{oe}R_L = \frac{55}{1 + (0.1 \times 10^{-6} \times 10,000)} = 54.9$$

The equation of Sec. 13-2 could also be used to solve the same problem. The equation is:

$$A_i = \frac{h_{21}Y_L}{h_{22} + Y_L}$$

where Y_L is load admittance. Although Y_L is a complex number, it can be considered as the reciprocal of load resistance. This would make $Y_L = 0.0001$.

Using the same 2N337 values:

$$\frac{55 \times 0.0001}{0.1 \times 10^{-6} + 0.0001} = 54.9$$

The y, z, and g parameters, although used to some extent in circuit design (especially y parameters), are not of great importance to the technician, and will not be discussed in detail. However, since it may be necessary for a technician to convert from one of these parameters to the more popular h parameter system, Sec. 13-2 contains the conversion data.

Another set of parameters, known as the s or *scattering* parameters, is used for design of high-frequency circuits. These s parameters are of little value to the technician since the s system is not used on most data sheets.

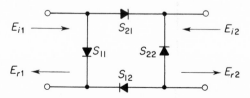

Fig. 13-5. Scattering parameters for linear two-port networks.

Figure 13-5 shows the signal flow diagram which forms the basis for s parameters, and the related equations.

13-2. Transister Parameter Conversion

13-2.1   Conversions among parameter types for y, z, h, and g parameters.

h to y

$$y_{11} = 1/h_{11} \quad y_{12} = -h_{12}/h_{11} \quad y_{21} = h_{21}/h_{11} \quad y_{22} = \Delta h/h_{11}$$

where $\Delta h = h_{11}h_{22} - h_{12}h_{21}$

y to h

$$h_{11} = 1/y_{11} \quad h_{12} = -y_{12}/y_{11} \quad h_{21} = y_{21}/y_{11} \quad h_{22} = \Delta y/y_{11}$$

where $\Delta y = y_{11}y_{22} - y_{12}y_{21}$

h to z

$$z_{11} = \Delta h/h_{22} \quad z_{12} = h_{12}/h_{22} \quad z_{21} = -h_{21}/h_{22} \quad z_{22} = 1/h_{22}$$

z to h

$$h_{11} = \Delta z/z_{22} \quad h_{12} = z_{12}/z_{22} \quad h_{21} = -z_{21}/z_{22} \quad h_{22} = 1/z_{22}$$

where $\Delta z = z_{11}z_{22} - z_{12}z_{22}$

h to g

$$g_{11} = h_{22}/\Delta h \quad g_{12} = -h_{12}/\Delta h \quad g_{21} = -h_{21}/\Delta h \quad g_{22} = h_{11}/\Delta h$$

g to h

$$h_{11} = g_{22}/\Delta g \quad h_{12} = -g_{12}/\Delta g \quad h_{21} = -g_{21}/\Delta g \quad h_{22} = g_{11}/\Delta g$$

where $\Delta g = g_{21}g_{22} - g_{12}g_{21}$

z to y

$$y_{11} = z_{22}/\Delta z \quad y_{12} = -z_{12}/\Delta z \quad y_{21} = -z_{21}/\Delta z \quad y_{22} = z_{11}/\Delta z$$

y to z

$$z_{11} = y_{22}/\Delta y \quad z_{12} = -y_{12}/\Delta y \quad z_{21} = -y_{21}/\Delta y \quad z_{22} = y_{11}/\Delta y$$

z to g

$$g_{11} = 1/z_{11} \quad g_{12} = -z_{12}/z_{11} \quad g_{21} = z_{21}/z_{11} \quad g_{22} = \Delta z/z_{11}$$

g to z

$$z_{11} = 1/g_{11} \quad z_{12} = -g_{12}/g_{11} \quad z_{21} = g_{21}/g_{11} \quad z_{22} = \Delta g/g_{11}$$

g to y

$$y_{11} = \Delta g/g_{22} \quad y_{12} = g_{12}/g_{22} \quad y_{21} = -g_{21}/g_{22} \quad y_{22} = 1/g_{22}$$

y to g

$$g_{11} = \Delta y/y_{22} \quad g_{12} = y_{12}/y_{22} \quad g_{21} = -y_{21}/y_{22} \quad g_{22} = 1/y_{22}$$

13-2.2   Conversions among common emitter, common base, and common collector parameters of the same type for y and h parameters.

Common emitter y parameters in terms of common base and common collector y parameters.

$$y_{11e} = y_{11b} + y_{12b} + y_{21b} + y_{22b} = y_{11c}$$
$$y_{12e} = -(y_{12b} + y_{22b}) = -(y_{11c} + y_{12c})$$
$$y_{21e} = -(y_{21b} + y_{22b}) = -(y_{11c} + y_{21c})$$
$$y_{22e} = y_{22b} = y_{11c} + y_{12c} + y_{21c} + y_{22c}$$

Common base y parameters in terms of common emitter and common collector y parameters.

$$y_{11b} = y_{11e} + y_{12e} + y_{21e} + y_{22e} = y_{22c}$$
$$y_{12b} = -(y_{12e} + y_{22e}) = -(y_{21c} + y_{22c})$$
$$y_{21b} = -(y_{21e} + y_{22e}) = -(y_{12c} + y_{22c})$$
$$y_{22b} = y_{22e} = y_{11c} + y_{12c} + y_{21c} + y_{22c}$$

Common collector y parameters in terms of common emitter and common base y parameters.

$$y_{11c} = y_{11e} = y_{11b} + y_{12b} + y_{21b} + y_{22b}$$
$$y_{12c} = -(y_{11e} + y_{12e}) = -(y_{11b} + y_{21b})$$
$$y_{21c} = -(y_{11e} + y_{21e}) = -(y_{11b} + y_{12b})$$
$$y_{22c} = y_{11e} + y_{12e} + y_{21e} + y_{22e} = y_{11b}$$

Common emitter h parameters in terms of common base and common collector h parameters.

$$h_{11e} = h_{11b}/(1 + h_{21b})(1 - h_{21b}) + h_{22b}h_{11b} \approx h_{11b}/(1 + h_{21b}) = h_{11c}$$
$$h_{12e} = h_{11b}h_{22b} - h_{12b}(1 + h_{21b})/(1 + h_{21b})(1 - h_{12b}) + h_{22b}h_{11b}$$
$$= 1 - h_{12c}$$
$$h_{12e} \approx (h_{11b}h_{22b})/(1 + h_{21b}) - h_{12b}$$
$$h_{21e} = -h_{21b}(1 - h_{12b}) - h_{22b}h_{11b}/(1 + h_{21b})(1 - h_{12b}) + h_{22b}h_{11b}$$
$$= -(1 + h_{21c})$$
$$h_{21e} \approx -h_{21b}/(1 + h_{21b})$$
$$h_{22e} = h_{22b}/(1 + h_{21b})(1 - h_{12b}) + h_{22b}h_{11b} \approx h_{22b}/(1 + h_{21b}) = h_{22c}$$

Common base h parameters in terms of common emitter and common collector h parameters.

$$h_{11b} = h_{11e}/(1 + h_{21e})(1 - h_{12e}) + h_{11e}h_{22e} \approx h_{11e}/1 + h_{21e}$$
$$h_{11b} = h_{11c}/h_{11c}h_{22c} - h_{21c}h_{12c} \approx -h_{11c}/h_{21c}$$
$$h_{12b} = h_{11e}h_{22e} - h_{12e}(1 + h_{21e})/(1 + h_{21e})(1 - h_{12e}) + h_{11e}h_{22e}$$
$$h_{12b} \approx (h_{11e}h_{22e}/1 + h_{21e}) - h_{12e}$$

$$h_{12b} = h_{21c}(1 - h_{12c}) + h_{11c}h_{22c}/h_{11c}h_{22c} - h_{21c}h_{12c}$$

$$h_{12b} \approx (h_{12c} - 1) - (h_{11c}h_{22c}/h_{21c})$$

$$h_{21b} = -h_{21e}(1 - h_{12e}) - h_{11e}h_{22e}/(1 + h_{21e})(1 - h_{12e}) + h_{11e}h_{22e}$$

$$h_{21b} \approx -h_{21e}/1 + h_{21e}$$

$$h_{21b} = h_{12c}(1 + {}_{21c}) - h_{11c}h_{22c}/h_{11c}h_{22c} - h_{21c}h_{12c} \approx -(1 + h_{21c}/h_{21c})$$

$$h_{22b} = h_{22e}/(1 + h_{21e})(1 - h_{12e}) + h_{11e}h_{22e} \approx h_{22e}/1 + h_{21e}$$

$$h_{22b} = h_{22c}/h_{11c}h_{22c} - h_{21c}h_{12c} \approx h_{22c}/h_{21c}$$

Common collector h parameters in terms of common base and common emitter h parameters.

$$h_{11c} = h_{11b}/(1 + h_{21b})(1 - h_{12b}) + h_{22b}h_{11b} \approx h_{11b}/1 + h_{21b} = h_{11e}$$

$$h_{12c} = 1 + h_{21b}/(1 + h_{21b})(1 - h_{12b}) + h_{22b}h_{11b} \approx 1 = 1 - h_{12e}$$

$$h_{21c} = h_{12b} - 1/(1 + h_{21b})(1 - h_{12b}) + h_{22b}h_{11b} \approx -1/1 + h_{21b}$$
$$= -(1 + h_{21e})$$

$$h_{22c} = h_{22b}/(1 + h_{21b})(1 - h_{12b}) + h_{22b}h_{11b} \approx h_{22b}/1 + h_{21b} = h_{22e}$$

13-2.3　Expressions for voltage gain, current gain, input impedance, and output impedance in terms of y, z, h, and g parameters.

Voltage gain:

$$A_v \text{ or } A_e = z_{21}Z_L/\Delta z + z_{11}Z_L$$
$$= -y_{21}/y_{22} + Y_L$$
$$= -h_{21}Z_L/h_{11} + \Delta h Z_L$$
$$= g_{21}Z_L/g_{22} + Z_L$$

where Z_L = load impedance, Y_L = load admittance
$$\Delta h = h_{11}h_{22} - h_{12}h_{21}$$

Current gain:

$$A_i = -z_{21}/z_{22} + Z_L$$
$$= -y_{21}Y_L/\Delta y + y_{11}Y_L$$
$$= h_{21}Y_L/h_{22} + Y_L$$
$$= -g_{21}/\Delta g + g_{11}Z_L$$

where $\Delta y = y_{11}y_{22} - y_{12}y_{21}$
$$\Delta g = g_{11}g_{22} - g_{12}g_{21}$$

Input impedance:

$$Z_{in} = \Delta z + z_{11}Z_L/z_{22} + Z_L$$
$$= y_{22} + Y_L/\Delta y + y_{11}Y_L$$
$$= \Delta h + h_{11}Y_L/h_{22} + Y_L$$
$$= g_{22} + Z_L/\Delta g + g_{11}Z_L$$

where $\Delta z = z_{11}z_{22} - z_{12}z_{21}$

Output impedance:

$$Z_{out} = \Delta z + Z_{22}Z_s/z_{11} + Z_s$$
$$= y_{11} + Y_s/\Delta y + y_{22}Y_s$$
$$= h_{11} + Z_s/\Delta h + h_{22}Z_s$$
$$= \Delta g + g_{22}Y_s/g_{11} + Y_s$$

where Y_s = source admittance, Z_s = source impedance

13-2.4 Conversion of s (scattering) and h parameters.

$$h_{11} = 1 - S_{11} + S_{22} - \Delta S/1 - S_{11} - S_{22} + \Delta S$$

$$h_{12} = 2S_{12}/1 - S_{11} + S_{22} - \Delta S$$

$$h_{21} = -2S_{21}/1 - S_{11} - S_{22} - \Delta S$$

$$h_{22} = 1 - S_{11} - S_{12} + \Delta S/1 - S_{11} + S_{22} - \Delta S$$

$$S_{11} = \Delta h + h_{11} - h_{22} - 1/\Delta h + h_{11} + h_{22} + 1$$

$$S_{12} = 2h_{12}/\Delta h + h_{11} + h_{22} + 1$$

$$S_{21} = -2h_{21}/\Delta h + h_{11} + h_{22} + 1$$

$$S_{22} = -\Delta h + h_{11} - h_{22} + 1/\Delta h + h_{11} + h_{22} + 1$$

where $\Delta h = h_{11}h_{22} - h_{12}h_{21}$
$\Delta S = S_{11}S_{22} - S_{12}S_{21}$

13-2.5 Hybrid h parameter relationships using *letter* subscripts.

Number Subscript	Common Emitter	Common Base	Common Collector
h_{11}	h_{ie}	$h_{ib} \approx h_{ie}/1 + h_{fe}$	$h_{ic} = h_{ie}$
h_{12}	h_{re}	$h_{rb} \approx (h_{ie}h_{oe}/1 + h_{fe}) - h_{re}$	$h_{rc} = 1 - h_{re}$
h_{21}	h_{fe}	$h_{fb} \approx -h_{fe}/1 + h_{fe}$	$h_{fc} = -(1 + h_{fe})$
h_{22}	h_{oe}	$h_{ob} \approx h_{oe}/1 + h_{fe}$	$h_{oc} = h_{oe}$

13-3. Determining Parameters at Different Frequencies

Transistor data sheets do not always specify all of the parameters needed. Likewise, the data sheets specify the parameters at some given operating frequency, when it is desired to know the parameter at another frequency. The following paragraphs describe methods for converting from one parameter to another at different frequencies.

Basic Parameter Relationships. Two major parameters are h_{fb} (Alpha, the common base a-c short-circuit forward current gain) and h_{fe} (Beta, the common emitter a-c short-circuit forward current gain). (Table 13-1 is a Glossary of the terms used in this paragraph.)

h_{fbo} (the value of h_{fb} at 1 kHz) will remain constant as frequency is increased, until a top limit is reached. After the top limit is reached, h_{fb} begins to drop rapidly. The frequency at which a significant drop in h_{fb} occurs provides a basis for comparison of the expected high frequency performance of different transistors. This frequency is known as $f_{\alpha b}$, and is defined at that frequency at which h_{fb} is 3 dB below h_{fbo}. Expressed in magnitude, h_{fb} at $f_{\alpha b}$ is 70.7 per cent of h_{fbo}. (Power gains, current gains, and voltage gains for a few common decibel values are found in Table 13-2.)

TABLE 13-1

Glossary of Parameter Symbols

Symbol	Definition
h_{fb}	Common base a-c forward current gain (Alpha)
h_{fbo}	Value of h_{fb} at 1 kHz
h_{fe}	Common emitter a-c forward current gain (Beta)
h_{feo}	Value of h_{fe} at 1 kHz
$f_{\alpha b}$	Common base current-gain cutoff-frequency. Frequency at which h_{fb} has decreased to a value 3 dB below h_{fbo}. ($h_{fb} = .707 h_{fbo}$)
$f_{\alpha e}$	Common emitter current-gain cutoff-frequency. Frequency at which h_{fe} has decreased to a value 3 dB below h_{feo}. ($h_{fe} = .707 h_{feo}$)
f_T	Gain bandwidth product. Frequency at which $h_{fe} = 1$ (0 dB)
G_{pe}	Common emitter power gain
f_{max}	Maximum frequency of oscillation. Frequency at which $G_{pe} = 1$ (0 dB)
$K\theta$	Excess phase-shifter factor. Factor which is a function of excess phase shift of current in the base of a transistor

TABLE 13-2

Conversion Table for Power, Voltage, and Current Ratios into Decibels

Decibels	Power Ratio	Voltage or Current Ratio
0	1.00	1.00
0.5	1.12	1.06
1.0	1.26	1.12
1.5	1.41	1.19
2.0	1.58	1.26
3.0	2.00	1.41
4.0	2.51	1.58
5.0	3.16	1.78
6.0	3.98	2.00
7.0	5.01	2.24
8.0	6.31	2.51
9.0	7.94	2.82
10.0	10.00	3.2
15.0	31.5	5.6
20.0	100.0	10.0
25	316.0	18.0
30	1,000.0	32.0
40	10,000.0	100.0
50	10^5	316
60	10^6	1,000.0

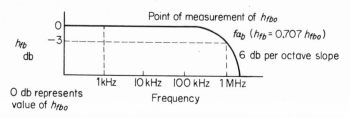

Fig. 13-6. Common base current gain versus frequency (*courtesy Motorola*).

A curve of h_{fb} versus frequency for a transistor with an $f_{\alpha b}$ of 1 MHz is shown in Fig. 13-6. This curve has the following significant characteristics:

(1) at frequencies below $f_{\alpha b}$, h_{fb} is nearly constant and approximately equal to h_{fbo};

(2) h_{fb} begins to decrease significantly in the region of $f_{\alpha b}$;

(3) above $f_{\alpha b}$, the rate of decrease in h_{fb} with increasing frequency approaches 6 dB per octave in the limit.

The curve of common base current gain versus frequency for any transistor has these characteristics, and the same general appearance as the curve of Fig. 13-6.

The *common emitter parameter* which corresponds to $f_{\alpha b}$ is $f_{\alpha e}$, the common emitter current-gain cutoff-frequency. $f_{\alpha e}$ is the frequency at which h_{fe} (Beta) has decreased 3 dB below h_{feo}. A typical curve of h_{fe} versus frequency for a transistor with an $f_{\alpha e}$ of 100 kHz is shown in Fig. 13-7. The curve of Fig. 13-7 has the same significant characteristics listed for Fig. 13-6.

These characteristics allow such a curve to be constructed for a particular transistor by knowing only h_{feo} and $f_{\alpha e}$. From the curve, h_{fe} at any frequency could be determined. Furthermore, if $f_{\alpha e}$ is not known, a curve could also be constructed if h_{feo} and h_{fe} at any frequency above $f_{\alpha e}$ were known. Thus, to find h_{fe} at any frequency, it is necessary to know only h_{feo} and either $f_{\alpha e}$ or h_{fe} at some frequency greater than $f_{\alpha e}$.

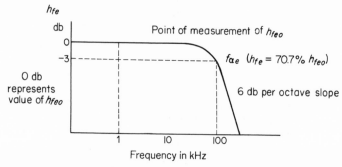

Fig. 13-7. Common emitter current gain versus frequency (*courtesy Motorola*).

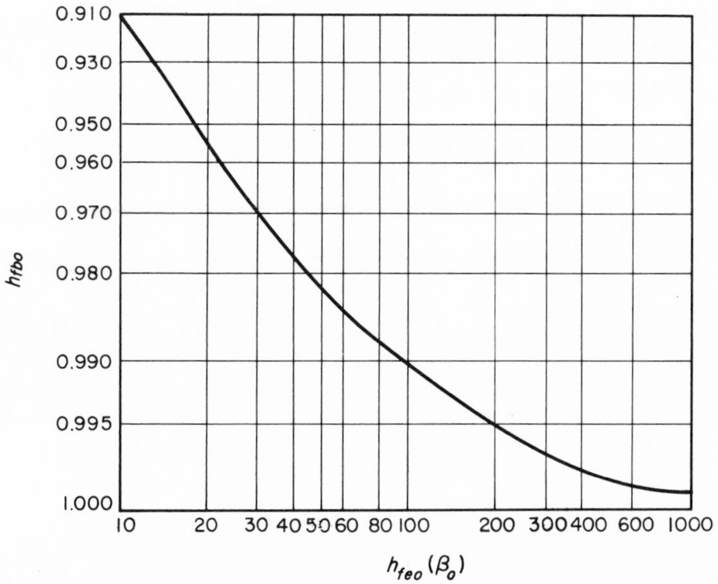

Fig. 13-8. Relationship of h_{feo} and h_{fbo} (courtesy *Motorola*).

Sometimes h_{feo} is needed and only h_{fbo} is given, or vice versa. The quantities h_{fbo} and h_{feo} are related by the following:

$$h_{feo} = h_{fbo}/1 - h_{fbo} \tag{13-1}$$

$$h_{fbo} = h_{feo}/h_{feo} + 1 \tag{13-2}$$

Equations 13-1 and 13-2 are plotted in Fig. 13-8. To simplify the problem, the low-frequency current-gain scales of Figs. 13-12 through 13-15 contain both an h_{fbo} and h_{feo} scale, and may be entered with a knowledge of either quantity.

Relationships between $f_{\alpha e}$ *and* $f_{\alpha b}$. Figures 13-12 through 13-15 provide a means of converting from $f_{\alpha e}$ to $f_{\alpha b}$, and vice versa. The nomograms in these illustrations are based on a $K\theta$ (excess phase-shift factor) of 0.5 to 1.0. Most transistors have a $K\theta$ in the 0.8 to 1.0 range. An exception is the germanium MADT types which have a $K\theta$ of 0.6. Alloy transistors have a $K\theta$ of 0.82.

The quantity $f_{\alpha e}$ is usually a much lower frequency than $f_{\alpha b}$ for the same transistor.

Determining h_{fe} at Different Frequencies

As previously stated, h_{fe} is probably the most important parameter, and is specified on almost all transistor data sheets. However, h_{fe} is usually specified for one frequency only. This limits the usefulness of the data sheet for practical purposes. Figures 13-8 through 13-15 can be used to determine h_{fe} at various frequencies. The following rules summarize how to find h_{fe} at a particular frequency.

(1) When the frequency is less than $f_{\alpha e}$, then $h_{fe} \approx h_{feo}$.

(2) When the frequency is equal to (or near) $f_{\alpha e}$, then $h_{fe} \approx 0.7h_{feo}$.

(3) When the frequency is greater than $f_{\alpha e}$, consider h_{fe} to be decreasing at 6 dB per octave at the frequency, and use Fig. 13-9 to find h_{fe}. The nomogram of Fig. 13-9 is based on entering the f-scale with the frequency

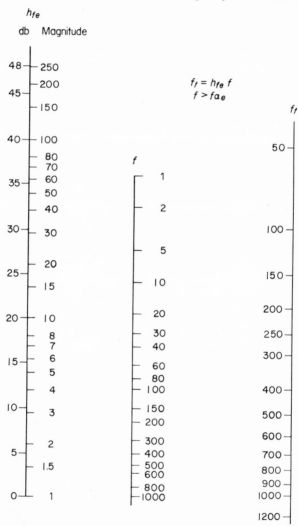

Fig. 13-9. Finding h_{fe} when frequency is greater than $f_{\alpha e}$ (courtesy Motorola).

at which it is desired to find h_{f_e}, and the f_T scale with the gain bandwidth product. f_T is defined as that frequency at which h_{f_e} is 1 (or 0 dB).

The value of f_T is sometimes specified indirectly on high frequency transistor data sheets. This is done by specifying h_{f_e} at some frequency above $f_{\alpha e}$. f_T is then obtained by multiplying the magnitude of h_{f_e} by the frequency of measurement. This relationship arises from the 6 dB per octave characteristic of the h_{f_e} versus frequency curve (Figs. 13-6 and 13-7) above $f_{\alpha e}$. Since 6 dB represent a current-gain magnitude of 2, h_{f_e} is halved each time frequency is doubled, and vice versa. Therefore, the product of h_{f_e} and the frequency of the sloping portion of the curve yields f_T.

f_T is also equal to the product of $h_{f_{eo}}$ and $f_{\alpha e}$, expressed by:

$$f_T = h_{f_{eo}} \times f_{\alpha e} \qquad (13\text{-}3)$$

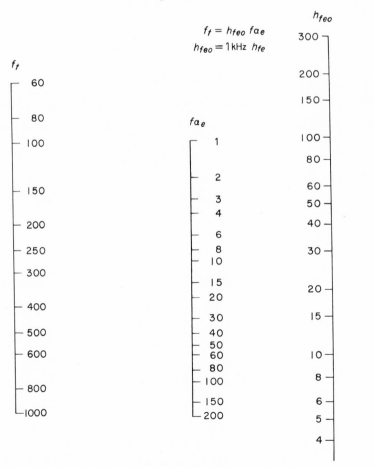

Fig. 13-10. Finding f_T when $f_{\alpha e}$ and $h_{f_{eo}}$ are known (*courtesy Motorola*).

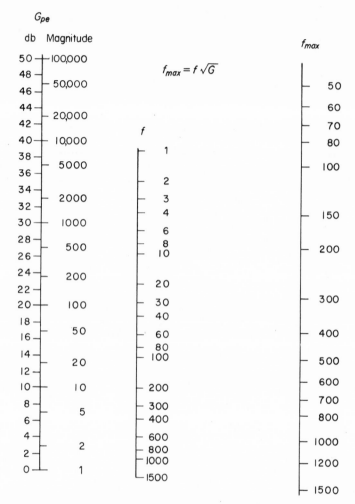

Fig. 13-11. Finding f_{max} when f and power gain are known (*courtesy Motorola*).

With h_{feo} known, equation 13-3 provides a simple means of finding $f_{\alpha e}$ when f_T is known, or vice versa. The equation 13-3 is represented by the nomogram of Fig. 13-10.

As a guide, remember that f_T is approximately equal to or slightly less than $f_{\alpha b}$ (the common base cutoff-frequency).

(4) If the data sheet specifies h_{fbo} instead of h_{feo}, use Fig. 13-8 to find h_{feo}.

If the data sheet does not specify f_T, use Fig. 13-10 to find f_T (with h_{feo} and $f_{\alpha e}$ known), or Fig. 13-13 to find f_T (with $f_{\alpha b}$ known). Use Fig. 13-15 instead of 13-13 to find f_T if the transistor is a MADT type.

If $f_{\alpha e}$ is not specified, use Fig. 13-10 to find $f_{\alpha e}$ (with f_T and h_{feo} known),

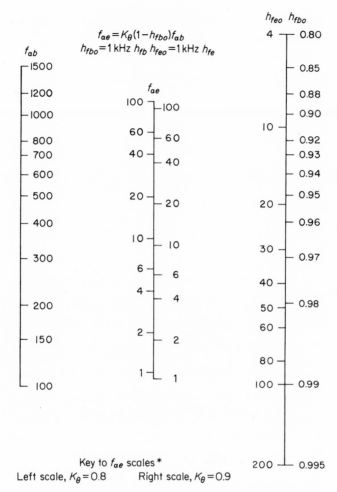

$$f_{ae} = K_\theta(1-h_{fbo})f_{ab}$$
$$h_{fbo} = 1\,\text{kHz } h_{fb} \quad h_{feo} = 1\,\text{kHz } h_{fe}$$

Key to f_{ae} scales *
Left scale, $K_\theta = 0.8$ Right scale, $K_\theta = 0.9$

Fig. 13-12. Finding f_{ae} when f_{ab} is known (*courtesy Motorola*).

or Fig. 13-12 to find f_{ae} (with f_{ab} known). Use Fig. 13-14 instead of Fig. 13-12 to find f_{ae} if the transistor is a MADT type.

Using the curves and nomograms. The following general instructions apply to the curves and nomograms of Figs. 13-6 through 13-15.

All power-gain and current-gain scales (except h_{fbo} and f_{feo}) are calibrated in both actual magnitudes (ratios) and decibel values for convenience.

The nomograms assume no shift in operating point. Known parameters used to find an unknown must be measured at the *same* collector voltage and collector current as the desired unknown.

Frequency scales on the nomograms are calibrated in numbers only, without units. Also, all nomograms contain two frequency scales. Decimal

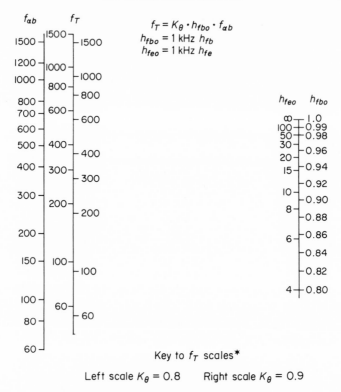

$$f_T = K_\theta \cdot h_{fbo} \cdot f_{\alpha b}$$
$$h_{fbo} = 1 \text{ kHz } h_{fb}$$
$$h_{feo} = 1 \text{ kHz } h_{fe}$$

Key to f_T scales*

Left scale $K_\theta = 0.8$ Right scale $K_\theta = 0.9$

Fig. 13-13. Relationship of f_T, $f_{\alpha b}$ and either h_{fbo} or h_{feo} (*courtesy Motorola*).

points may be shifted on the frequency scales of any nomograms as long as they are shifted the same amount on both scales. That is, both frequency scales of a nomogram must be multiplied by 10 to the same power. This permits the same nomogram to be used for both high- and low-frequency transistors.

The nomograms assume that both power gain and current gain decrease with increasing frequency at a rate of 6 dB per octave at high frequencies.

Example of using the nomograms. The following is an example of how the nomograms can be used to find various parameters of a transistor. The data sheet of a General Electric 2N337 transistor shows a "typical" current transfer ratio h_{fe} of 55 (magnitude) and a "typical" Alpha cutoff-frequency of 30 MHz. (The data sheet lists Alpha cutoff-frequency as f_{hfb}, rather than $f_{\alpha e}$.) Assume that it is desired to find the h_{fe} at 2.5 MHz. Also assume that the transistor has a phase shift factor $K\theta$ of 0.9 (approximate).

Using Fig. 13-13 with a known $f_{\alpha b}$ of 30 MHz and a h_{fbo} (h_{fe} at 1 kHz) of 55, we find a f_T of 28 MHz (approximate).

Using Fig. 13-9 with a f_T of 28 MHz, and the desired frequency f of 2.5

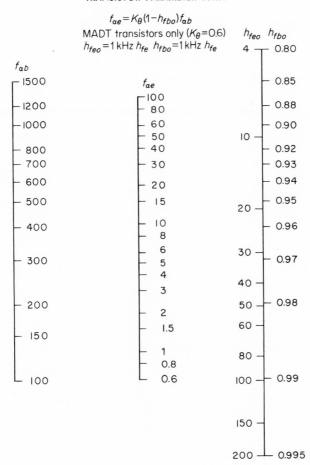

$$f_{ae} = K_\theta(1 - h_{fbo})f_{ab}$$

MADT transistors only ($K_\theta = 0.6$)
$h_{feo} = 1\,\text{kHz}\,h_{fe}$ $h_{fbo} = 1\,\text{kHz}\,h_{fe}$

Fig. 13-14. Finding f_{ae} when f_{ab} is known (for MADT type transistors)
(courtesy Motorola).

MHz, we find a h_{fe} of approximately 12 (magnitude) or 23 dB. This figure is close to the "typical" 24 dB figure of h_{fe} at 2.5 MHz, also shown on the data sheet).

Using Fig. 13-10 with f_T of 28 MHz and h_{fbo} of 55, we find f_{ae} to be 0.5 MHz.

As an alternate method, we could have used Fig. 13-12 to find f_{aem}, using the known f_{ab} and h_{fbo}. Once f_{ae} had been found, Fig. 13-10 could have been used to find f_T.

Determining maximum operating frequency of a transistor. Although common emitter current gain is equal to 1 at f_T, there may still be considerable power gain at f_T due to different input and output impedance levels. Thus, f_T is not necessarily the highest useful frequency of operation for a

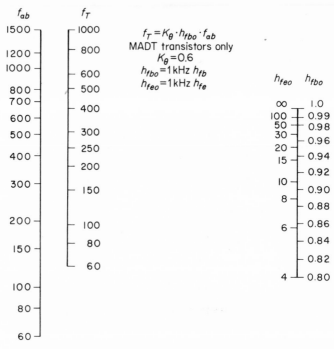

Fig. 13-15. Relationship of f_T, $f_{\alpha b}$ and either h_{fbo} or h_{feo} (for MADT type transistors). (*courtesy Motorola*)

transistor. An additional parameter, the maximum frequency of oscillation ($f_{\max}$), is found.

The term $f_{\max}$ is the frequency at which common emitter *power* gain is equal to 1, and is related to f_T by:

$$f_{\max} \approx \sqrt{\frac{f_T}{25.12 \times r_b' \times C_c}}$$

where r_b' is base resistance

C_c is collector capacitance

A plot of common emitter power gain versus frequency has the same characteristics as shown in Figs. 13-6 and 13-7. This leads to another gain bandwidth product:

$$f_{\max}\sqrt[f]{\text{power gain}}$$

where f is the frequency of measurement, and power gain is expressed in *magnitude*, not in decibels.

$f_{\max}$ may be found by measuring power gain at some frequency on the 6 dB per octave portion of the power gain versus frequency curve, and multiplying the square root of the power gain with the frequency of measurement.

The nomogram of Fig. 13-11 permits the f_{max} to be found when the common emitter power gain (G_{pe}) is known (either in magnitude or dB).

The parameters are voltage and current dependent, and operating point must be considered in all cases. For example, the high-frequency h_{fe} measured at one collector voltage and current must not be used to calculate f_T directly at another voltage and/or current without considering the added effects of the different operating point.

The parameter $f_{\alpha e}$ for present high-frequency transistors usually lies in the region between 10 and 50 MHz. The term h_{fe}, measured at any frequency above this region, is assumed on the 6 dB per octave portion of h_{fe} versus frequency curve, and is used to calculate f_T directly.

Power gain measured at any frequency above 50 MHz is assumed on the 6 dB per octave portion of the power gain versus frequency curve, and is used to calculate f_{max} directly.

13-4. Determining Parameters at Different Temperatures

The following paragraphs describe methods for determining transistor parameters at temperatures other than those specified on the data sheet.

Collector-base leakage current I_{cbo} versus temperature. Figure 13-16 shows the typical change of I_{cbo} with temperature for a number of different power transistors. Five typical curves (three germanium and two silicon) are shown. The dotted lines indicate the increase in I_{cbo} (rule of thumb). For germanium

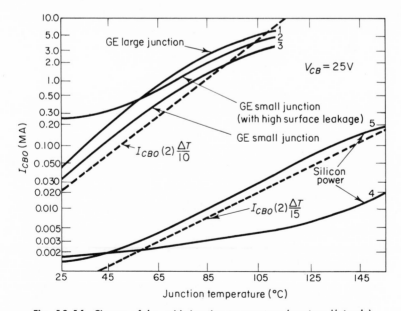

Fig. 13-16. Change of I_{cbo} with junction temperature (*courtesy Motorola*).

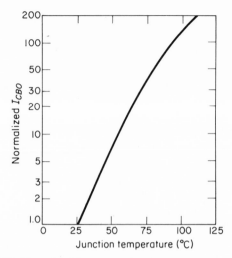

Fig. 13-17. Normalized I_{cbo} versus junction temperature ($V_{cb}=2V$) (*courtesy Motorola*).

transistors, the I_{cbo} doubles every 10°C increase in temperature. For silicon transistors, the I_{dbo} doubles every 15°C increase.

Figure 13-17 shows a much closer prediction of the change of I_{cbo} with temperature. This is a normalized I_{cbo} curve with a V_{cb} of 2 volts, and is based on the similarity of the slopes of the I_{cbo} curves for germanium transistors as indicated by the low surface-leakage devices of Fig. 13-16.

Current gain versus temperature. Figure 13-18 shows the variation of collector current versus base current for a germanium power transistor over a temperature range of −40°C to +100°C. The dotted line indicates the actual base current readings obtained at 100°C. However, because I_{cbo} is flowing through the base in the opposite direction, we must combine the I_{cbo} current with the meter reading (subtract I_{cbo} from I_b) to get the actual gain component of base current, as shown by the solid line. This solid line curve

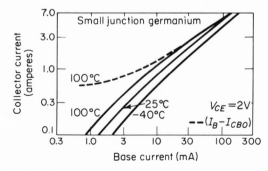

Fig. 13-18. Collector current versus base current (*courtesy Motorola*).

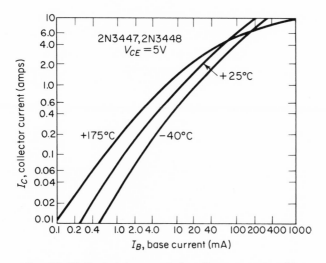

Fig. 13-19. Typical I_c versus I_b for silicon power transistors
(*courtesy Motorola*).

is more useful in determining the increase in transistor-gain characteristics with temperature because the solid line eliminates the highly variable I_{cbo} component. The curve of Fig. 13-18 shows that transistor gain increases approximately 50 per cent (rule of thumb) when the temperature is raised from 25°C to 100°C (assuming class *B* operation).

The gain of the transistor also increases as the voltage across it is increased, and is directly related to the output impedance of the transistor. Because output impedance will change with temperature, the same germanium transistor will show approximately 100 per cent increase in gain at a voltage level equal to $1/2\ BV_{ceo}$ (collector-emitter breakdown) over the same temperature increase of 25°C to 100°C. If such information is not stated on the transistor data sheet, the curve of Fig. 13-18 is *approximately* true, and can be used for most germanium power transistors operated as class *B*.

Figure 13-19 shows the variation of collector current versus base current for a typical diffused silicon power transistor over the temperature span of −40°C to +175°C. The curve of Fig. 13-19 shows that for silicon power transistors (operated as class *B*) there will be an approximate increase of 100 per cent in current gain when the temperature is raised from 25°C to 175°C. At higher voltages, the change will be somewhat greater. However, the effect of output impedance changes on silicon transistors will be less than on germanium transistors.

Base-to-emitter voltage V_{be} versus temperature. Figure 13-20 shows the transconductance (base voltage versus collector current) curve versus temperature of a typical germanium power transistor operating as class *B*.

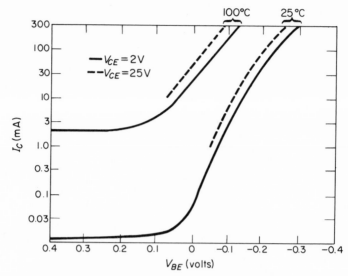

Fig. 13-20. Transconductance versus temperature (with V_{BE} equal to -0.2 to $+0.8$) *(courtesy Motorola)*.

The solid curves show the change in V_{be} (at a V_{ce} of 2 volts) from 25°C to 100°C. In the range of 30 milliamperes to 300 milliamperes collector current, V_{be} changes at the rate of approximately 2 millivolts/°C.

Although there is a change in V_{be} with an increase in collector voltage (as shown by the dotted lines of Fig. 13-20), the slope of the V_{be} curve is generally not greatly affected at 25°C over the collector current ranges that apply to class *B* operation. However, if the slope of the V_{be} curve is changed at higher temperatures because of a temperature effect on the output impedance at high voltage levels, this effect must be considered. The dotted lines of Fig. 13-20 show the change in V_{be} with a change in temperature for a high-output impedance transistor when operating at a collector voltage equal to half the BV_{ceo} rating. This indicates that the rate of change for this transistor is increased at 2.2 mv/°C. Such would be a useful temperature coefficient for V_{be} operating as class *B*.

Figure 13-21 shows the effects of temperature on the transconductance of a typical diffused silicon power transistor. The curves of Fig. 13-21 show that V_{be} changes 2.2 mv/°C over a temperature range of 25°C to 175°C in class *B* operation. This change in V_{be} with temperature is shown at a collector voltage of 5 volts. Because of the very high output impedance of silicon power transistors operating as class *B*, 2.2 mv/°C would be valid except at very high voltage levels. This may not be true with class *A* operation (due to higher current), depending upon transistor characteristics.

Thermal resistance. The thermal resistance of junction-to-ambient (θ_{JA}) is an important characteristic in design of transistor circuits. This is especially

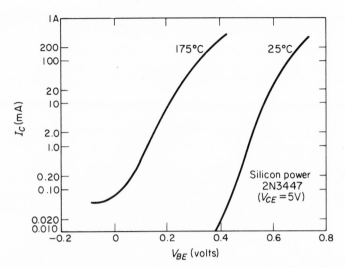

Fig. 13-21. Transconductance versus temperature (*courtesy Motorola*).

true in *power* transistors. Usually, power transistor data sheets specify some value of thermal resistance at a given temperature. For each increase in temperature from this specified value, there will be a change in the temperature-dependent characteristics of the transistor. Since there will be a change in temperature with changes in *power dissipation* of the transistor, the junction-to-ambient temperature will also change, resulting in a characteristic change. Therefore, the transistor characteristics can change with ambient temperature changes, and with changes produced by variation in power dissipation.

When heat sinks are used, the long thermal-time constant of most heat sinks will cause the transistor characteristics to be slow. The thermal resistance of both the *transistor and heat sink* should be known if thermal stability is to be properly evaluated.

Manufacturers use many different ways to specify thermal resistance. No matter what method is used, thermal resistance can be defined as the *increase in temperature of the transistor junction* (*with respect to some reference*) *divided by the power dissipated.*

In power transistors, the thermal resistance is normally measured from the transistor pellet to the case, resulting in the term θ_{JC}. On those transistors where the case is bolted directly to the mounting surface with an integral threaded bolt or stud, the terms θ_{MB} (thermal resistance to mounting base) or θ_{MF} (thermal resistance to mounting flange) are used.

These terms take into consideration only the thermal paths from junction-to-case (or mount). For power transistors in which the pellet is mounted directly on the header or pedestal, the total internal thermal resistance from junction-to-case (or mount) varies from 50°C per watt to less than 1°C per watt. If a transistor is not mounted on a heat sink, the thermal resistance from case-to-ambient air θ_{CA} is so large in relation to that from junction to case (or mount) that the total thermal resistance from junction-to-ambient air θ_{JA} is primarily the result of the θ_{CA} term.

The following Table 13-3 lists values of case-to-ambient air thermal resistances for a number of popular transistor cases. From this table it will be seen that the large, heavy-duty cases such as TO-3 will have a small temperature increase (for a given wattage) in comparison to such cases as the TO-5. That is, the heavy-duty cases will dissipate the heat into the ambient air.

After about 1 watt (or less) it becomes impractical to increase the size of the case to make the case-to-ambient term comparable to the junction-to-case term. For this reason, most power transistors are designed for use on an external heat sink. Sometimes, the chassis or mounting area serves as the heat sink. In other cases, a heat sink is attached to the case. Either way, the primary purpose of the heat sink is to increase the effective heat-dissipation area of the case, and provide a low heat resistance path from case to ambient.

The thermal resistance of a heat sink can be calculated if the following factors are known: material, mounting position, exact dimensions, shape, thickness, surface finish, color. Even if all of these factors are known, the thermal resistance calculations are approximate. From a practical stand-point, it is better to accept the manufacturer's specifications for a heat sink. The heat sink thermal resistance actually consists of two series elements: the thermal resistance from case-to-heat sink that results from conduction (θ_{CS}), and the thermal resistance from heat sink to ambient air caused by convection and radiation (θ_{SA}).

To operate a transistor at its full power capabilities, there should be no temperature difference between the case and ambient air. This will occur only when the thermal resistance of the heat sink is zero, and the only thermal resistance is that between junction and case. It is not practical to manufacture a heat sink with zero resistance. However, the greater the ratio

TABLE 13-3

Case-to-Ambient Thermal Resistance for
Typical Transistor Cases

Case	$\theta_{CA}(°C/W)$
TO-3	30
TO-5	150
TO-8	75
TO-18	300
TO-36	25
TO-39	150
TO-46	300
TO-60	70
TO-66	60

TABLE 13-4

Comparison of Insulating Washers Used for Electrical Isolation
of Transistor TO-3 Case from Heat Sink

Material	Thickness (Inches)	θ_{CS} (°C/W)	Capacitance (pf)
Beryllia	0.063	0.25	15
Anodized aluminum	0.016	0.35	110
Mica	0.002	0.4	90

of θ_{JC}/θ_{CA}, the nearer the maximum power limit (set by θ_{JC}) can be approached.

When transistors are to be mounted on heat sinks, some form of electrical isolation must be provided between the case and the heat sink (unless a grounded-collector circuit is used). Because good electrical insulators usually are also good thermal insulators, it is difficult to provide electrical insulation without introducing some thermal resistance between case and heat sink. The best materials for this application are mica, beryllium oxide (Beryllia), and anodized aluminum. A comparison of the properties of these three materials for case-to-heat sink isolation of a TO-3 case is shown in Table 13-4.

For small, general-purpose transistors with a TO-5 case, a beryllium-oxide washer can be used to provide insulation between case and a metal chassis or printed-circuit board. The use of a zinc-oxide-filled silicone compound (such as Dow Corning #340 or Wakefield #120) between the washer and chassis, together with a moderate amount of pressure from the top of the transistor, helps to decrease thermal resistance. If the transistor is mounted

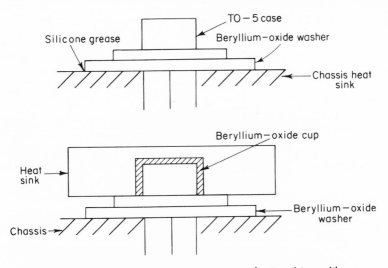

Fig. 13-22. Typical mounting arrangements for transistors with TO-5 package (*courtesy RCA*).

within a heat sink, a beryllium cup should also be used between the transistor and the heat sink. Figure 13-22 shows both types of mounting.

Commercial fin-type heat sinks are available for various transistor case sizes and shapes. Such heat sinks are especially useful when the transistors are mounted in Teflon sockets which provide no thermal conduction to the chassis or printed-circuit board. Commercial heat sinks are rated by the manufacturer as to the thermal resistance, usually in terms of °C/watt. When heat sinks involve the use of washers, the °C/W factor usually includes the thermal resistance between case and sink, or θ_{cs}.

All other factors being equal, the heat sink with the lowest thermal resistance (°C/W) is best. That is, a heat sink with 1°C/W is better than a 3°C/W heat sink. Of course, the heat sink must fit the transistor case, and the space around the transistor. Except for these factors, selection of a suitable heat sink should be no particular problem.

One exception to this is when the transistors are to be operated at high frequencies. Any isolation between collector and the chassis (produced by the washer between case and heat sink) will also result in capacitance between the two metals. This capacitance can be a problem, especially at frequencies above 100 MHz, and must be considered in r-f circuit design.

Relationship of Thermal Resistance and Dissipation Capability of Transistors. The main problem in operating a transistor at or near its maximum power dissipation limits is a condition known as *thermal runaway.* When current passes through a transistor junction, heat will be generated. If *all* of this heat is not dissipated by the case (which is not possible), the junction temperature will increase. This, in turn, will cause more current to flow through the junction, even though the voltage, circuit values, etc., remain the same. In turn, this will cause the junction temperature to increase even further, with a corresponding increase in current flow. If the heat is not dissipated by some means, the transistor will burn out and be destroyed.

The remainder of this section is devoted to calculations of mechanical heat dissipation.

As in the case of other characteristics, transistor manufacturers specify maximum power dissipation in a variety of ways on their data sheets. Some manufacturers provide *safe operating area curves* for temperature and/or power dissipation of transistors. Other manufacturers specify a maximum power dissipation, in relation to a given ambient temperature, or a given case temperature. Still others specify a maximum junction temperature, or a maximum case temperature.

If a transistor is operated under *steady-state conditions*, the maximum dissipation capability is dependent upon three factors: the sum of the series thermal resistances from the transistor junction to ambient air, the maximum junction temperature, and the ambient temperature. The relationship between these three factors is as follows:

Assume that it is desired to find the maximum power dissipation of a transistor (in watts). The following conditions are specified: a maximum junction temperature of 200°C, a junction-to-case thermal resistance (θ_{JC}) of 2°C/W, a heat sink with a thermal resistance (θ_{SA}) of 3°C/W, an ambient temperature of 25°C. Note that the thermal resistance of the heat sink includes any thermal resistance produced by the washer between the transistor case and heat sink. (If this factor were not known, it is a good rule-of-thumb to add 0.5°C/W thermal resistance for any washer between case and heat sink. As shown in Table 13-4, this thermal resistance value is high, and allows a safe tolerance.)

First, find the total junction-to-ambient thermal resistance θ_{JA}:

$$\theta_{JA} = \theta_{JC} + \theta_{SA} \quad \text{or} \quad 5 = 2 + 3$$

Next, find the maximum permitted power dissipation:

$$\frac{\text{Maximum}}{\text{power}} = \frac{\text{Maximum junction temperature} - \text{ambient temperature}}{\theta_{JA}}$$

or

$$\frac{200 - 25}{5} = 35 \text{ watts maximum}$$

The case temperature of a transistor operated under these conditions would be:

$$T_C = \text{maximum power} \times (\theta_{SA}) + \text{ambient} \quad \text{or}$$

$$T_C = (35 \times 3) + 25 = 130°C \text{ case temperature}$$

Now assume that a maximum case temperature had been specified, rather than a maximum junction temperature. In that event, subtract the ambient temperature from the maximum permitted case temperature:

$$130° - 25° = 105°C$$

Then divide the case temperature by the heat sink thermal resistance:

$$\frac{105°C}{3°C} = 35 \text{ watts maximum power}$$

When a transistor is operated by a *single, nonrepetitive pulse*, the maximum permitted power dissipation is much greater than with steady-state operation just described. To obtain maximum permitted power under single pulse conditions, the *transient thermal resistance* must be calculated. Usually, the transient thermal resistance is shown on the safe operating area curves in the form of a *power multiplier*. The power multiplier is given for a specific case temperature, and given pulse width. For example, a 2N3055 transistor has power multipliers of 2.1 for 100-millisecond pulses, 3.0 for 1-millisecond pulses, and on up to to 7.7 for 30-microsecond pulses. The steady-state (or direct current) maximum power is multiplied by these factors to obtain the maximum single pulse power. For example, if the maximum permitted steady state power is 100 watts for a given set of conditions, the single pulse maximum power would be 300 watts if a 1-millisecond pulse were used.

Usually, the data sheets will specify the power multipliers for a given case temperature. If the case temperature is increased, the factor must be derated. The temperature derating factor is obtained by:

$$\text{Derating factor} = \frac{\text{Case temperature} - \text{ambient}}{\begin{array}{l}\text{Maximum}\\ \text{junction} \quad - \text{ambient}\\ \text{temperature}\end{array}}$$

For example, assume that a transistor has a maximum junction temperature of 200°C, that the case temperature under these conditions is 130°C, that the ambient temperature is 25°C, and that the maximum steady-state power is 35 watts. Further assume that the transistor data sheet specifies a power multiplier of 3.0 for 1-millisecond pulses, with a case temperature of 25°C.

The derating factor would be:

$$\frac{130°C - 25°C}{200°C - 25°C} = \frac{3}{5}$$

To obtain the maximum single pulse power:

$$\begin{array}{l}\text{Maximum}\\ \text{power}\end{array} = \text{Multiplier} \times (1 - \text{derating factor}) \times \text{steady-state power}$$

or

$$3 \times \left(1 - \frac{3}{5}\right) \times 35 = 42 \text{ watts}$$

These calculations for single pulse operation are based on the assumption that the heat sink capacity is large enough to prevent the heat sink temperature from rising between pulses.

When a transistor is operated by *repetitive pulses*, the heat sink and case temperature will rise. This must be taken into account when determining the maximum power dissipation. The maximum permitted power dissipation for a transistor operated with repetitive pulses is calculated by:

$$\begin{array}{l}\text{Maximum}\\ \text{permitted}\\ \text{power}\end{array} = \cfrac{\begin{array}{l}\text{Power}\\ \text{multiplier}\end{array} \times \left(\begin{array}{l}\text{maximum}\\ \text{junction} \quad - \text{ambient}\\ \text{temperature}\end{array}\right)}{\theta_{JC} + \begin{array}{l}\text{Power}\\ \text{multiplier}\end{array} \times \left(\begin{array}{l}\text{duty}\\ \text{cycle}\\ \text{percentage}\end{array}\right)\theta_{JA}}$$

Assume that it is desired to find the maximum permitted power of the same transistor previously described for steady-state operation, but now operated by 1-millisecond pulses repeated at 100 Hz. The conditions are:

power multiplier 3.0, θ_{JC} 2°C/W, θ_{JA} 5°C/W (including heat sink), maximum junction temperature 200°C, ambient temperature 25°C, and a duty cycle of 10 per cent (1-millisecond on/10-milliseconds off).

$$\text{Maximum permitted power} = \frac{3(200 - 25)}{2 + 3(0.1)5} = 150 \text{ watts}$$

The case temperature of a transistor operated by repetitive pulses would be:

$$\text{Case temperature} = \begin{matrix} \text{Peak} \\ \text{pulse} \\ \text{power} \end{matrix} \times \begin{matrix} \text{Duty} \\ \text{cycle} \\ \text{percentage} \end{matrix} \times \theta_{JA} + \text{ambient}$$

Peak pulse power is obtained by multiplying the collector voltage by the collector current (assuming that the transistor is operated in a grounded-emitter or grounded-base configuration, and that the transistor is switched full-on and full-off by the pulses).

For example, assume an ambient temperature of 25°C, a duty cycle of 10 per cent, a total θ_{JA} of 5°C/W, and peak pulses of 120 watts (say a collector voltage of 60, and a collector current of 2 amperes). The case temperature would be:

$$\text{Case temperature} = 120 \times 0.1 \times 5 + 25 = 85°\text{C}$$
$$(\text{or } T_C)$$

CHAPTER 14

Integrated Circuit Data

Typically, an integrated circuit, or IC, consists of transistors, resistors, and diodes, mounted on or etched into a semiconductor material. The material is usually silicon, and in the form of a "chip" or wafer. All of the parts are interconnected (by techniques similar to those used in printed circuit boards) to perform a definite circuit function or operation. It must therefore be realized that the IC concept is one of a complete (or nearly complete) circuit, rather than a group of related semiconductor devices. However, to make the IC package an operable unit, it must still be connected to a power source, an input and an output. In most cases, the output must also be connected to external components such as capacitors and coils, since it is not practical to combine these parts on the semiconductor chip.

It is the purpose of this chapter to introduce the reader to ICs, and to describe (in practical terms) how an IC package can be connected to form an operable unit. It is not intended that this publication be considered as an IC design handbook. However, the data in this chapter will be of great value to the laboratory technician working in the IC field.

14-1. Packaging and Internal Construction

In theory, an integrated circuit semiconductor chip could be connected directly to the power source, input, etc. However, this is not practical in most cases due to the very small size of the chip. IC chips are almost always microminiature. Instead of direct connection, the chip is mounted in a

| 8-lead | 10-lead | 12-lead | Flat pack | Dual in-line |

Fig. 14-1. Typical integrated circuit packaging (*courtesy* RCA).

suitable container, and connected to the external circuit through leads on the container.

There are three basic schools of thought on IC packaging: the transistor package, the flat pack, and the dual in-line package. Some typical examples of these are shown in Fig. 14-1.

In the transistor package, the chip is mounted inside a transistor case such as the TO-5. Instead of the usual three leads (emitter, collector, and base), there will be 8 or 10 leads to accommodate the various power source and input/output connections required in a complete circuit.

In the flat pack, the chip is encapsulated in a rectangular case, with terminal leads extending through the sides and ends.

Although there has been some attempt at standardization for IC terminal connections, the various manufacturers still use their own systems. It is therefore necessary to consult the data sheet for the particular IC when making connections from an external circuit.

14-2. Circuit Differences

Although the basic circuits used in ICs are similar to those of transistors, there are certain differences. For example, inductances (coils) are almost never found as part of an IC. This is because it is next to impossible to form an inductance on the same material as the transistors and resistors. Likewise, large value capacitors (above about 30 to 60 pF) are not formed as part of the IC. When a large value capacitor, or an inductance of any type, is a necessary part of a circuit, these components are part of the external circuit.

Transistors are often used in place of resistors in IC construction. Usually, such a transistor is a FET (refer to Chapter 7), since the basic field effect transistor acts somewhat like a resistor. Figure 14-2 shows how a FET can be substituted for a resistor in an IC. In this circuit, the FET gate is returned to one side of the supply. With such an arrangement, the FET takes up less space than a corresponding resistor, and provides a much higher power dissipation capability.

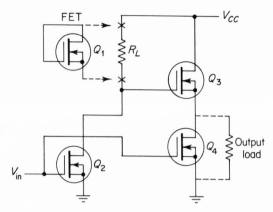

Fig. 14-2. How a FET can be substituted for a resistor in an IC.

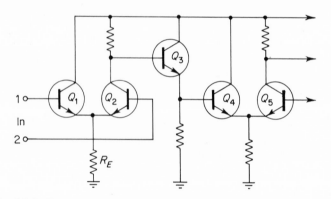

Fig. 14-3. Transistor Q_3 substitutes for a coupling capacitor in typical IC.

ICs often use direct-coupled circuits to eliminate capacitors. Figure 14-3 shows how a transistor Q_3 is used to eliminate the need for a capacitor between $Q_1 - Q_2$ and $Q_4 - Q_5$. By eliminating the capacitor, the frequency range of the circuit is also extended.

14-3. Basic IC Types

There are two basic types of ICs; *digital* and *linear*.

Digital ICs

Digital ICs are the integrated circuit equivalents of basic transistor logic circuits. Like their transistor counterparts, digital ICs are used in computers, digital telemetry, etc., and form such circuits as gates, counters, choppers, multivibrators, shift registers, etc. It must be remembered that a digital IC is a complete functioning logic network, usually requiring nothing more than an input, output, and power source.

The following paragraphs describe a group of typical digital ICs.

The *direct-coupled transistor logic* (DCTL) of Fig. 14-4 shows three possible combinations where the transistors are connected *directly*. That is, diodes, resistors, or other transistors are not used for coupling.

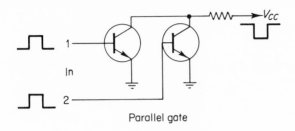

Parallel gate

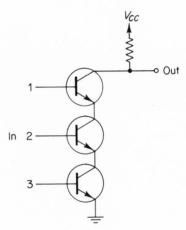

Series gate

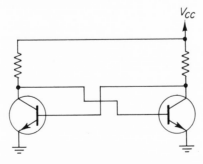

Basic flip—flop

Fig. 14-4. Direct-coupled transistor logic (DCTL).

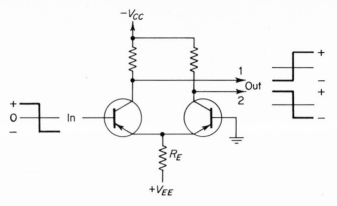

Fig. 14-5. Current-mode logic (CML).

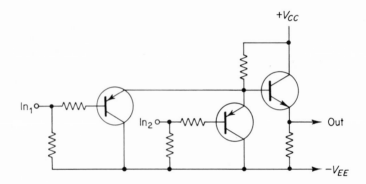

Fig. 14-6. Complementary transistor logic (CTL).

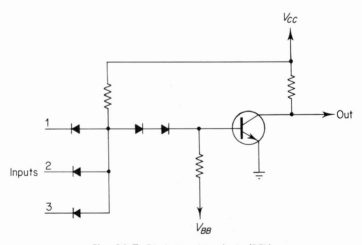

Fig. 14-7. Diode-transistor logic (DTL).

The *current-mode logic* (CML) of Fig. 14-5 shows a split-load differential amplifier with complementary outputs. Sometimes this circuit is found single-ended with the second transistor omitted, and the emitter of the first transistor clamped to ground through a diode.

The *complementary transistor logic* (CTL) of Fig. 14-6 shows complementary transistors in the input (*PNP*) and output (*NPN*). By using an

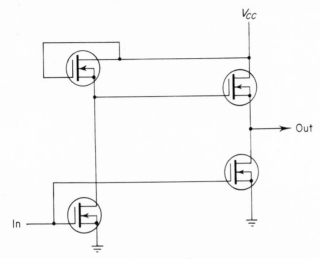

Fig. 14-8. Metal-oxide-semiconductor transistor logic (MOSTL).

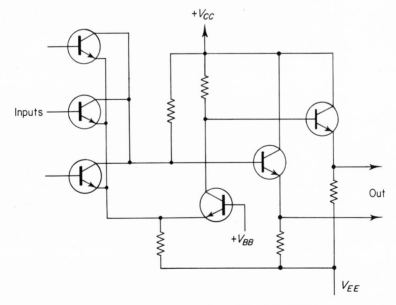

Fig. 14-9. Emitter-coupled transistor logic (ECTL).

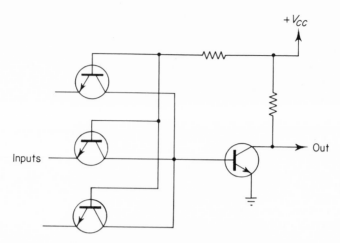

Fig. 14-10. Transistor-transistor logic (TTL).

emitter-follower output, many outputs can be taken from the circuit without serious loading effects.

The *diode-transistor logic* (DTL) of Fig. 14-7 shows how diodes and transistors can be combined to form typical NOR/NAND circuits. A NOR circuit for negative-going input and NAND circuit for positive-going input are shown.

The *metal-oxide-semiconductor transistor logic* (MOSTL) of Fig. 14-8 shows a logic circuit using FETs throughout. Such a circuit has high input impedance, and a multiple output capability.

The *emitter-coupled transistor logic* (ECTL) of Fig. 14-9 shows a circuit similar to the CML of Fig. 14-5, except that multiple inputs are used, and the circuit operates in the nonsaturated mode. The base of Q_2 is biased positive so that Q_2 will always conduct, even though the inputs are off.

The *transistor-transistor logic* (TTL) of Fig. 14-10 shows a circuit similar to the DTL of Fig. 14-7, except the gate diodes are replaced by transistors. In the circuit of Fig. 14-10, it is possible to set the input threshold voltage over a wide range.

Linear ICs

Linear ICs are the integrated circuit equivalents of basic transistor circuits such as amplifiers, oscillators, mixers, frequency multipliers, modulators, limiters, detectors, and so on. Although linear ICs are complete functioning circuits, they often require additional external components (in addition to a power supply) for satisfactory operation. Typical examples of such external components are: a resistor to convert a linear amplifier into an operational amplifier, a resistor-capacitor combination to provide frequency compensation, and a coil-capacitor combination to form a filter (for band pass or band rejection).

The *differential amplifier* is the most common type of linear IC. Figure 14-11 shows a typical circuit (courtesy Fairchild). As shown, the complete circuit is contained in a TO-5 package with 8 terminal leads.

The basic purpose of such a circuit is to produce an output signal that is linearly proportional to the *difference between two signals applied to the input*. The circuit shown will provide an over-all open-loop gain of approximately 2500.

As shown, the inputs are applied to two emitter-coupled transistor amplifiers Q_1 and Q_2. The collectors of Q_1 and Q_2 are connected to a constant current source, and have the same value load resistor. The output signals from Q_1 and Q_2 are equal, but 180° out-of-phase. Should only one input be used (with the other input grounded or returned to a bias), the resulting output will still appear at both collectors (Q_1 and Q_2) but the gain will be cut in half.

The *operational amplifier* is another common form of linear IC. The operational amplifier is a high-gain direct-coupled circuit where the *gain and frequency* response are controlled by external feedback networks. It is possible to convert a differential amplifier such as shown in Fig. 14-11 to an

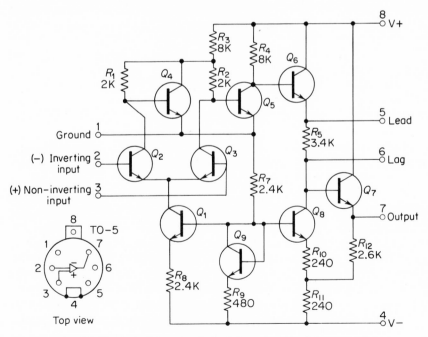

Fig. 14-11. Typical integrated circuit housed within 8-terminal TO-5 package (*courtesy Fairchild*).

operational amplifier by the addition of one feedback resistor (as is discussed in later sections of this chapter).

The operational amplifier is used in low-level instrumentation, analog computers, servo systems, and wherever controlled gain and frequency response are necessary.

14-4. Selection of External Components for Integrated Circuits

Since ICs can be used to form almost endless types of operational units (amplifiers, filters, oscillators, mixers, digital logic, etc.), there are many design equations to determine the values for components used with ICs. The subject is so complex that a complete treatment (even from the non-technical approach) is beyond the scope of this publication.

However, one of the most common problems is the conversion of a differential amplifier into an operational amplifier. This involves connecting a power source, selecting a feedback resistance of correct value, and possibly selecting frequency compensation components of correct value. These steps will be covered in the following paragraphs, using the differential amplifier circuit of Fig. 14-11 as an example.

It should be noted that the equations in the following paragraphs are simplifications of much longer equations, and ignore the fact that input impedance and open-loop gain are always less than infinity, and that output impedance is never zero.

Connecting the Power Source

Most ICs require connections to *both a positive and negative* power source. A few ICs can be operated from a single power supply source. Many ICs require equal power supply voltages (such as +9 volts and −9 volts). However, this is not the case with the example circuit of Fig. 14-11, which requires a +9 volts at pin 8 and a −4.2 volts at pin 4.

Unlike most transistor circuits, where it is common to label one power supply lead positive and the other negative without specifying which (if either) is common or ground, it is *necessary that all IC power supply voltages be referenced to a common or ground.* Manufacturers do not agree on power supply labeling for ICs. For example, the Fairchild IC of Fig. 14-11 uses $V+$ to indicate the positive voltage, and $V-$ to indicate the negative voltage. Another manufacturer might use the symbols V_{EE} and V_{CC} to represent negative and positive, respectively. As a result, the IC data sheet should be studied carefully before applying any power source.

No matter what labeling is used, the IC will require two power sources with the positive lead of one and the negative lead of the other tied to ground. A typical arrangement is shown in Fig. 14-12.

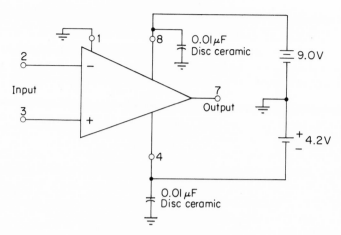

Fig. 14-12. Typical power supply connections for an IC (*courtesy Fairchild*).

The two capacitors shown in Fig. 14-12 provide for decoupling of the power supply. Usually, disc ceramic capacitors are used. The capacitors should always be connected as close to the IC pins as possible, not at the power supply terminals.

It should be noted that the case (TO-5) of the Fig. 14-11 circuit is connected to pin 4. Therefore, the case will be below ground (or "hot") by 4.2 volts.

Connecting the Feedback Resistance

As in the case of any feedback circuit, the output is connected to the input through a resistance. The differential amplifier of Fig. 14-11 has two inputs. One input is labeled (−) and is inverting. That is, a signal applied to this input will be inverted (out-of-phase) at the output. The other input is labeled (+) and is noninverting. That is, a signal applied to this input will remain in-phase at the output.

The feedback from the output must always be connected to the (−) inverting input. This will cause the out-of-phase to partially cancel some of the input, and reduce over-all gain. This will also stabilize the output when the input signal is varied in frequency, and will extend the (flat response) frequency range of the amplifier.

If the feedback were connected to the (+) noninverting input, the in-phase output will add to the input, and cause oscillation. When a differential amplifier is used as an operational amplifier, the (+) noninverting input should be grounded, or returned to ground through a resistance. As a rule of thumb, the ground resistance for the (+) noninverting input should be approximately equal to (or slightly less than) the source resistance (or impedance) connected to the (−) inverting input.

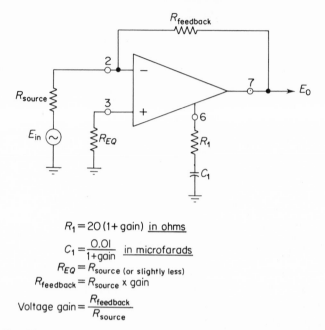

$$R_1 = 20\,(1 + \text{gain})\ \underline{\text{in ohms}}$$
$$C_1 = \frac{0.01}{1 + \text{gain}}\ \underline{\text{in microfarads}}$$
$$R_{EQ} = R_{\text{source}}\ \text{(or slightly less)}$$
$$R_{\text{feedback}} = R_{\text{source}} \times \text{gain}$$
$$\text{Voltage gain} = \frac{R_{\text{feedback}}}{R_{\text{source}}}$$

Fig. 14-13. Typical connections for converting a differential IC into an operational amplifier.

Figure 14-13 shows the usual feedback connection arrangement, as well as the equations for determining voltage gain, and the values of the frequency compensation RC network.

Note that the voltage gain is equal to the feedback resistance, divided by the input or source resistance (or impedance). Assuming a source resistance of 10K, the voltage gain would be 10 if the feedback resistance were 100K. Therefore, to find the feedback resistance value, multiply the source resistance by the desired voltage gain.

For example, assume a 1K source resistance, and a desired voltage gain of 100. The correct feedback resistance would be 100K.

It should be noted that the equation of Fig. 14-13 is for closed-loop gain (with feedback). This gain can never be higher than the open-loop gain (without feedback).

It should also be noted that not all linear ICs will have the same frequency compensation network as shown in Fig. 14-13. Some linear ICs have no provision for external frequency compensation components. Therefore, the equations do not apply in all cases.

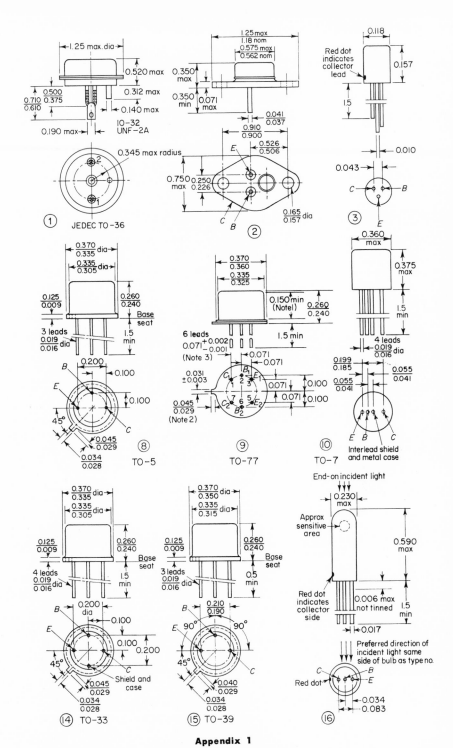

Appendix 1

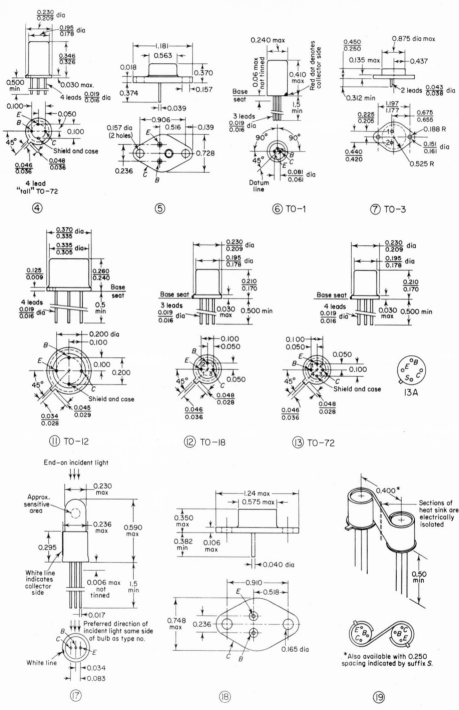

④
4 lead "tall" TO-72

⑤

⑥ TO-1

⑦ TO-3

⑪ TO-12

⑫ TO-18

⑬ TO-72

13A

⑰

⑱

⑲

*Also available with 0.250 spacing indicated by suffix S.

Appendix 2

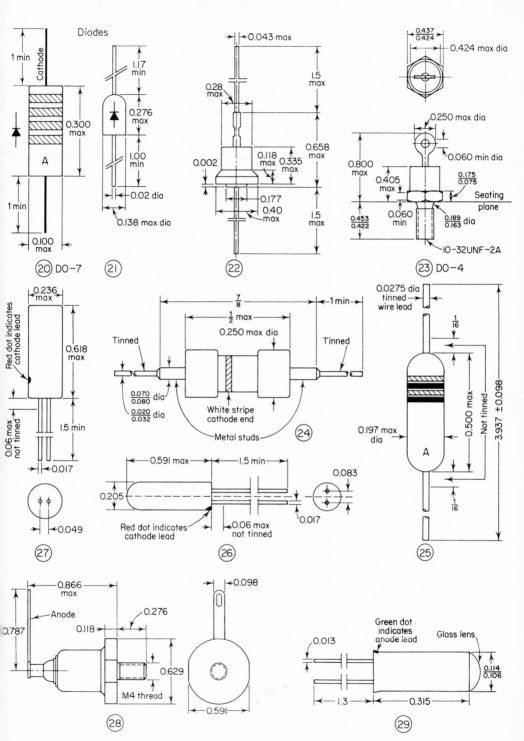

Diodes

20 DO-7 21 22 23 DO-4

24

25 26 27

28 29

Appendix 3

Diodes (cont.)

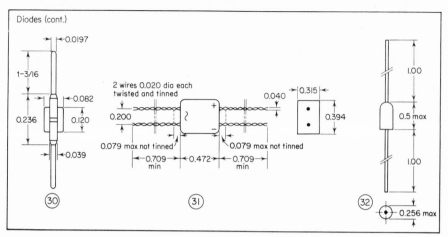

Subminiature

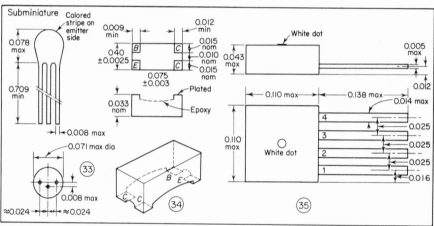

Heat sinks

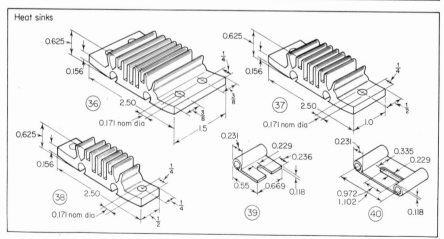

Appendix 4

TYPICAL TRANSISTOR SPECIFICATION 2N332

Voltages:

Collector to base (emitter open)	V_{CBO}	45 volts
Emitter to base (collector open)	V_{EBO}	1 volt

Collector current I_C 25 ma

Power*

Collector dissipation (25°C.)	P_C	150 mw
Collector dissipation (125°C.)	P_C	50 mw

Temperature range:

Storage	T_{STG}	−65°C. to 200°C.
Operating	T_A	−55°C. to 175°C.

ELECTRICAL CHARACTERISTICS (25°C.)
(Unless otherwise specified, $V_{CB} = 5$ v; $I_E = -1$ ma; $f = 1$ kc)

Small signal characteristics:		min.	nom.	max.	
Current transfer ratio	h_{fe}	9	15	20	
Input impedance	h_{ib}	30	53	80	ohms
Reverse voltage transfer ratio	h_{rb}	.25	1.0	5.0	$\times 10^{-4}$
Output admittance	h_{ob}	0.0	.25	1.2	μmhos
Power gain					
($V_{CE} = 20$ v; $I_E = -2$ ma; $f = 1$ kc;					
$R_G = 1$ K ohms; $R_L = 20$ K ohms)	G_e		35		db
Noise figure	NF		28		db

High frequency characteristics:					
Frequency cutoff					
($V_{CB} = 5$ v; $I_E = -1$ ma)	f_{ab}		15		mc
Collector to base capacity					
($V_{CB} = 5$ v; $I_E = -1$ ma; $f = 1$ mc)	C_{ob}		7		μμf
Power gain (common emitter)					
($V_{CB} = 20$ v; $I_E = -2$ ma; $f = 5$ mc)	G_e		17		db

D-c characteristics:					
Collector breakdown voltage					
($I_{CBO} = 50$ μa; $I_E = 0$; $T_A = 25$°C.)	BV_{CBO}	45			volts
Collector cutoff current					
($V_{CB} = 30$ v; $I_E = 0$; $T_A = 25$°C.)	I_{CBO}		.02	2	μa
($V_{CB} = 5$ v; $I_E = 0$; $T_A = 150$°C.)	I_{CBO}			50	μa
Collector saturation resistance					
($I_B = 1$ ma; $I_C = 5$ ma)	R_{SC}		80	200	ohms

Switching characteristics:					
($I_{B_1} = 0.4$ ma; $I_{B_2} = -0.4$ ma;					
$I_C = 2.8$ ma)					
Delay time	t_d		.75		μsec
Rise time	t_r		.5		μsec
Storage time	t_s		.05		μsec
Fall time	t_f		.15		μsec

*Derate 1mw/°C increase in ambient temperature.

Appendix 5

Index